Interpreting Organic Spectra

Interpreting Organic Spectra

David Whittaker

Department of Chemistry, University of Liverpool, UK

ROYAL SOCIETY OF CHEMISTRY

ISBN 0-85404-601-1

A catalogue record for this book is available from the British Library.

Published by The Royal Society of Chemistry,
Thomas Graham House, Science Park, Milton Road, Cambridge CB4 0WF, UK

For further information see our web site at www.rsc.org

Typeset by Paston PrePress Ltd, Beccles, Suffolk, NR34 9QG
Printed and bound by Athenaeum Press Ltd, Gateshead, Tyne & Wear

Preface

This book, like many other text-books, has its origins in a course given by the author. The course was for beginners in spectroscopy; it was a conventional series of lectures on techniques, with tutorials spent interpreting spectra. This quickly proved that while lecturing is a good way of teaching spectroscopy, lectures on how to interpret spectra have something in common with lectures on how to ride a bicycle. Since most organic chemists regard spectroscopy as a means of gaining structural information about compounds in which they are interested, the course was modified to become a series of workshops, in which a short lecture on a technique was followed by many examples to interpret, while post-graduate students helped with early difficulties. This resulted in a big improvement in the ability of students to get structural data out of spectra, and made the post-graduates wish that they had learned spectroscopy this way, but the exercises in combining the results of all the techniques were less successful. This book aims to correct this problem and is the course I would choose to teach if sufficient workshops were available. It tries to show that each technique has a particular speciality, and that these need to be combined to identify a structure which is consistent with all the spectra.

In a course like this, the question of how much to write about a technique is debatable. One extreme would be to consider only the spectra, and to ignore how they are produced; the other is to delve too deeply into the theory of the method, and leave the student uncertain what to look for in the actual spectrum. I have aimed to be closer to the first extreme than the second, since some excellent detailed books on spectroscopy are available. The theory parts of this book cover approximately the same ground as the video tapes on IR, MS, UV, ^{13}C and ^{1}H spectra which were made at the University of Liverpool and which are available from the Royal Society of Chemistry.

The spectra should also be useful to teachers who are not trying to teach all the areas of spectroscopy covered here. In many cases, the mass spectrum can be replaced by a molecular formula, though this substantially increases the difficulty of a few problems. Many ^{13}C problems can be solved without ^{1}H, and *vice versa*, though this will make a few problems too difficult for beginners.

Compiling a collection of spectra like this has required a lot of help, and I am particularly grateful for the help of Sandra Hedges, Ann Leyden, Steve Apter and Allan Mills in running most of them. Thanks are also due to Frank Doran, who discovered samples in remote corners of the Chemistry Department, and Dr. J. A. Race of Micromass, who was very helpful with mass spectrometry. Finally, I must thank Frances Poole, for typing the manuscript.

David Whittaker

Contents

CHAPTER 1

Infrared Spectroscopy

Infrared spectroscopy is the best means of identifying functional groups in a molecule. It involves measuring the absorption by a substance of radiation in the region from 4000 to $600 \, \text{cm}^{-1}$. The range covers roughly 2.5–$16 \, \mu\text{m}$ ($1 \, \mu\text{m} = 10^{-4} \, \text{cm}$), but the frequency scale is now used universally.

Every bond in a molecule vibrates, resulting in a change in its dipole moment. This change in dipole moment provides a mechanism for the absorption of radiation. The energy of the vibration is such that the radiation absorbed is in the infrared region, which is of lower frequency, and hence lower energy, than visible light. Consequently, every bond in a molecule has an absorption peak in the infrared spectrum or the Raman spectrum of the molecule. Every substance therefore has its own unique infrared spectrum, so that we can identify any organic material by comparing its infrared spectrum with that of a known sample. In addition, each different functional group, such as O–H, C–H or C=C, absorbs within a narrow range of frequencies so that we can identify a functional group present in a molecule by the presence of an absorption band in a particular range of the infrared spectrum.

The frequency at which a bond absorbs radiation depends on the masses of the atoms forming the bond. The bonds which absorb radiation at the upper end of the frequency range are those which involve a light atom, hydrogen, with a heavier atom, such as nitrogen, oxygen or carbon. Hydrogen, being light, vibrates strongly and rapidly, so we see a strong, high energy absorption. As we move to lower frequencies, we come to the vibrations of two heavier atoms, such as C–N, C–O and C–C, and finally, at the lowest energies, we find the vibrations of the heavier atoms bonded to very heavy atoms, such as C–Cl, C–Br and C–I. Within this general trend, multiple bonds absorb higher energy radiation than single bonds, so the C≡C bond absorbs at higher frequency than the C=C bond.

Before we study an infrared spectrum, we should consider how it is obtained. The sample to be studied is usually examined in one of four ways:

(1) As a pure liquid.
(2) As a mull. This method is used for solids. The solid is ground finely using a pestle and mortar, and mixed with a small amount of a liquid hydrocarbon (liquid paraffin) and run as a liquid. This has the disadvantage that the spectrum of the liquid hydrocarbon is superimposed on the spectrum of the sample.
(3) The compound is finely ground, mixed with KBr, compressed and run as a disc. This avoids adding extra absorption, but is time consuming.
(4) The compound is dissolved in a suitable solvent, and run as a solution. The spectrum cannot be observed in regions where the solvent absorbs.

Now let us consider an infrared spectrum. The spectrum of phenylamine, $PhNH_2$ (also known as aniline), is shown in Figure 1.1.

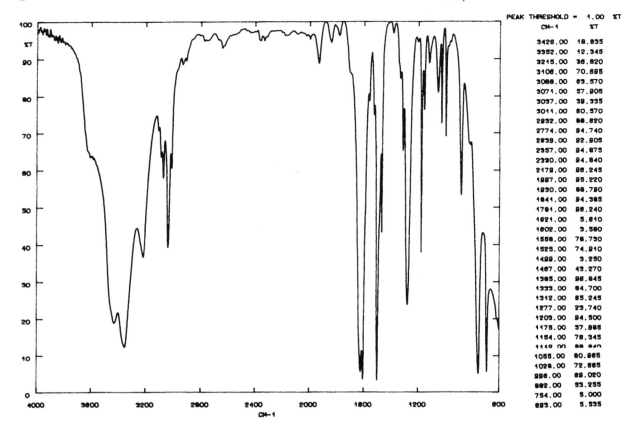

PEAK THRESHOLD =	1.00 %T
CM-1	%T
3426.00	18.835
3352.00	12.345
3215.00	36.820
3106.00	70.695
3086.00	83.570
3074.00	57.905
3037.00	39.335
3011.00	80.570
2932.00	88.620
2774.00	94.740
2639.00	92.905
2357.00	94.875
2330.00	94.840
2179.00	96.245
1997.00	95.220
1930.00	88.790
1841.00	94.385
1791.00	96.240
1621.00	5.610
1602.00	3.580
1556.00	78.730
1525.00	74.910
1499.00	3.250
1467.00	43.270
1385.00	96.845
1333.00	84.700
1312.00	85.245
1277.00	23.740
1203.00	94.500
1176.00	37.995
1154.00	78.345
1119.00	88.840
1055.00	80.985
1029.00	72.885
996.00	89.020
882.00	53.255
754.00	5.000
693.00	5.535

Figure 1.1

The spectrum records the amount of radiation transmitted at each frequency, so the maximum absorption occurs when the least light is transmitted, and the recorder line is closest to the bottom of the spectrum. Positions of maximum absorption are difficult to measure accurately from the spectrum, so most spectrometers record them automatically, and print them alongside the spectrum, giving the % of radiation transmitted (%T) alongside. The machine can be set to record frequencies of all or only the stronger peaks.

Phenylamine has an $-NH_2$ group, which characteristically absorbs in the region from 3500 to 3250 cm^{-1}. The spectrum actually has two peaks in this range, at 3426 and 3352 cm^{-1}. These do NOT represent a peak for each N–H bond (the bonds are indistinguishable so must have identical absorption frequencies): they result from the in-phase and out-of-phase vibrations of the N–H bonds:

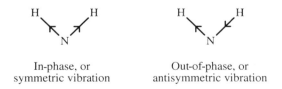

In-phase, or   Out-of-phase, or
symmetric vibration   antisymmetric vibration

If the N–H bond is part of an amide, hydrogen bonding moves the absorption to lower frequency. If we have a secondary amine, with only one N–H bond, we obtain a single peak, which can easily be confused with the O–H bond.

A genuine O–H bond is shown in the next spectrum, that of butan-2-ol, $CH_3CH_2CH(OH)CH_3$. This is shown in Figure 1.2. The O–H bond absorbs

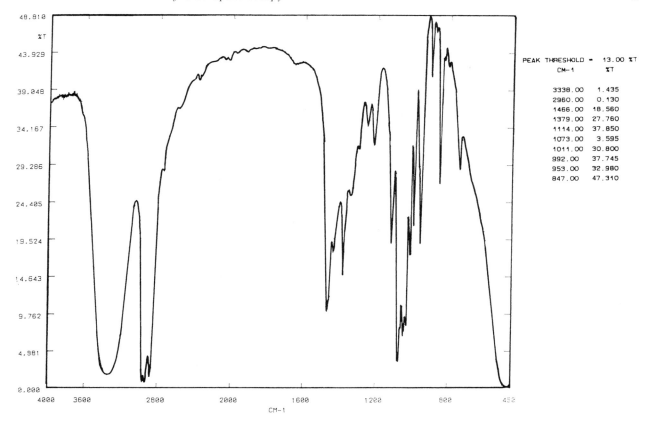

PEAK THRESHOLD = 13.00 %T

CM-1	%T
3338.00	1.435
2960.00	0.130
1466.00	18.560
1379.00	27.760
1114.00	37.850
1073.00	3.595
1011.00	30.800
992.00	37.745
953.00	32.980
847.00	47.310

Figure 1.2

radiation in the range 3700–3200 cm^{-1}, and we find that butan-2-ol has a peak at 3338 cm^{-1}. The O–H peak, like the N–H peak but unlike other peaks in the spectrum, is a broad rounded peak rather than the usual sharp spiked peaks. This is a result of intermolecular hydrogen bonding. The hydrogen atom on any particular oxygen atom is probably attached to another oxygen atom by hydrogen bonding, so bond vibrations vary over a frequency range, and the broad peak which we see is an envelope covering many absorptions at slightly different frequencies. If we dilute the sample of butan-2-ol with solvent which cannot form a hydrogen bond, such as dichloromethane, we get the spectrum shown in Figure 1.3, in which the broad O–H bond peak is reduced, and a small sharp absorption peak has appeared at higher frequency, resulting from the non-hydrogen bonded O–H bond. At greater dilution of the alcohol dissolved in dichloromethane, the original O–H peak has almost vanished, and the non-hydrogen bonded peak has further increased in size, as shown in Figure 1.4.

The O–H bond of a carboxylic acid shows even stronger hydrogen bonding than that of an alcohol, since the acid is so strongly hydrogen bonded as to exist in the dimeric form in the pure liquid:

$$
\begin{array}{c}
\text{O} \cdots \text{H} - \text{O} \\
\text{R} - \text{C} \diagup \qquad \diagdown \text{C} - \text{R} \\
\text{O} - \text{H} \cdots \text{O}
\end{array}
$$

As a result, the carboxylic acid O–H group has an extremely broad absorption peak in the range 3200–2200 cm^{-1}. Combined with a carbonyl peak in the range 1725–1680 cm^{-1}, this makes identification of a carboxylic acid from its

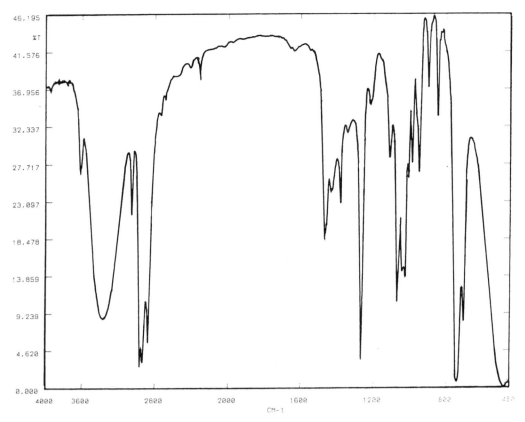

Figure 1.3

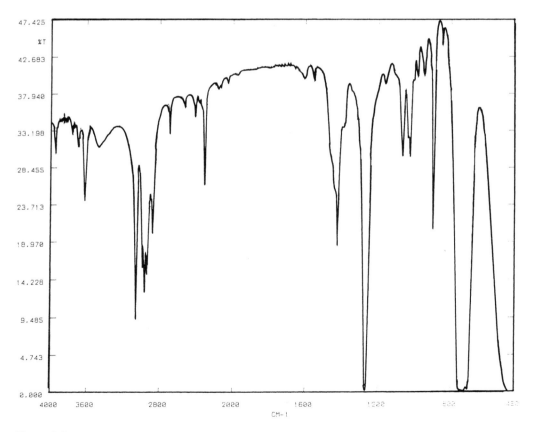

Figure 1.4

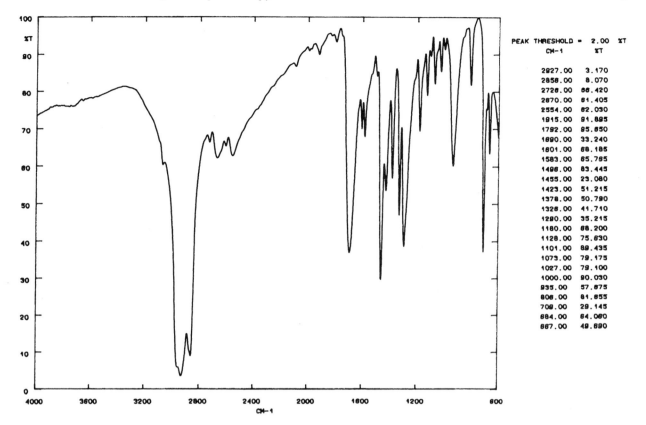

Figure 1.5

infrared spectrum an easy task. A typical carboxylic acid spectrum is that of benzoic acid, PhCOOH, which is shown in Figure 1.5.

As we move to lower frequencies in the infrared range, the next peak that we encounter is the C–H peak in the region 3100–2700 cm^{-1}. This supplies little useful structural information since almost all organic compounds contain hydrogen, and in the case of a solid run as a mull it will certainly be present in the liquid hydrocarbon used to make the mull. After the C–H absorption, we leave the high-frequency high-intensity X–H bonds, and, after a usually blank region of the spectrum, encounter the weaker triple bond absorption frequencies. The most common are the C≡N bond, which absorbs at 2300–2200 cm^{-1}, and the C≡C bond, which absorbs in the region from 2250 to 2100 cm^{-1}. This latter bond is particularly weak if neither carbon atom is bonded to hydrogen. The carbonyl group occurs next, absorbing over the unusually wide range of 1850–1640 cm^{-1}. However, the variation of the position of the C=O peak within this range is very valuable diagnostically, and will be considered separately.

As we further descend the frequency scale, we next come to the C=C bond region, which covers 1680–1620 cm^{-1} in unconjugated systems, and can go as low as 1590 cm^{-1} in conjugated systems. The C=C bond is often of low intensity, especially when fully substituted, and it is easily missed. In an aromatic system, the C–C bonds usually have two or three absorption peaks between approximately 1600 and 1500 cm^{-1}, usually with the strongest peak at about 1500 cm^{-1}. The spectrum of benzoic acid shown in Figure 1.5 illustrates this with weak peaks at 1601 cm^{-1} and 1583 cm^{-1}, and a stronger peak at 1496 cm^{-1}.

The region below about 1500 cm^{-1} is usually complex, and contains relatively little information about functional groups. It is often referred to as

Table 1.1 *Infrared absorption frequencies of common functional groups*

Functional group	Absorption range (cm^{-1})
N–H	3500–3250
CON–H	3500–3000
O–H	3700–3200
COO–H	3200–2200
C–H	3100–2700
C≡N	2300–2200
C≡C	2250–2100
C=O	1850–1640
C=C	1680–1590
Aromatic	1600–1500

the 'fingerprint region', since this is the region of greatest value when comparing the spectrum of an unknown compound with that of a known compound. At the lower frequency end of this region we come to the C–Cl bond absorption in the region 800–400 cm^{-1}. The C–Br and C–I bonds usually absorb at too low a frequency for most spectrometers. The halogens are more readily identified by mass spectrometry.

The absorbing frequencies discussed above have been collected together in Table 1.1.

The carbonyl absorption was referred to earlier as covering a very wide range of frequencies. The absorption frequency is unusually sensitive to substitution at the carbonyl carbon atom. The reason for this is that the carbonyl group exists as a resonance hybrid of two forms:

$$C{=}O \leftrightarrow C^{+}{-}O^{-}$$

The two forms absorb at different frequencies, the polar $C^{+}{-}O^{-}$ having the higher frequency absorption. If we now consider the hybrid

it is easy to see that electron withdrawing substituents will favour the $C^{+}{-}O^{-}$ form, and electron supplying substituents will favour the C=O form.

When the substituent X is a chlorine atom, this strongly electron withdrawing group favours the $C^{+}{-}O^{-}$ bond, so acid chlorides absorb in the region around 1815–1790 cm^{-1}. The carbonyl group is another electron withdrawing substituent, so that the anhydride absorbs in this region, but since we have two carbonyl groups close together we can have in-phase and out-of-phase vibrations, as with the NH$_2$ group. Saturated carboxylic acid anhydrides thus show two peaks, one in the region 1850–1800 cm^{-1} and the other in the region 1790–1740 cm^{-1}.

The position of the carbonyl peak, however, does not depend entirely on inductive effects. In esters, the overlap of the lone pair of electrons on X with the C=O bond reduces the double bond character of the C=O bond, so the ester carbonyl group absorbs in the range 1755–1735 cm^{-1}. Aldehydes and ketones come next at 1740–1700 cm^{-1}, then the carboxylic acid carbonyl absorbs at 1725–1700 cm^{-1}, which is lower than might have been expected, but its position is affected by the dimeric nature of the carboxylic acid group in solution.

All these examples have referred to the carbonyl carbon atom attached to a

Table 1.2 *Infrared absorption frequencies of substituted carbonyl groups*

	Absorption frequency range (cm^{-1})	
Functional group	*R = Aliphatic*	*R = Conjugated alkene or aromatic*
Anhydride, RCOOCOR	1850–1800 and 1790–1740	1830–1775 and 1770–1710
Acid chloride, RCOCl	1815–1790	1790–1750
Ester, RCOOR*	1800–1745	
Ester, RCOOAlkyl	1755–1730	1735–1710
Aldehyde, RCOH	1740–1715	1715–1680
Ketone, RCOAlkyl	1725–1700	1700–1665
Carboxylic acid, RCOOH	1725–1700	1715–1680
Amide, RCON	1680–1640	1695–1665

*R = aryl or vinyl.

saturated group; if it is attached to a multiple bond or an aromatic ring, then the carbonyl absorption frequency is lowered by 15–40 cm^{-1}. Thus, if we need to assign a carbonyl group, it is first necessary to see if the molecule shows signs of unsaturation; if this is next to the carbonyl carbon atom, then carbonyl frequencies will be lower than expected. Unexpectedly, conjugation shifts the amide carbonyl to higher frequency. It may be due to an inductive effect on the conjugated CO–N system.

The carbonyl absorption frequencies have been collected together in Table 1.2.

Problems in Interpreting Infrared Spectra

Using the data given in Tables 1.1 and 1.2, you should now be able to identify the common functional groups present in the compounds whose spectra comprise Samples 1 to 20. Remember that it is not sufficient to identify a compound as simply a carbonyl compound. If you observe a carbonyl peak, you should check for the presence of an aromatic ring or a double bond in the molecule, then use the data in Table 1.2 to say what type of carbonyl group (*e.g.* anhydride, ketone) is present in the sample. Note that some spectra are consistent with more than one structural type, since ranges of frequencies can overlap. All the spectra are run on pure liquids or nujol mulls.

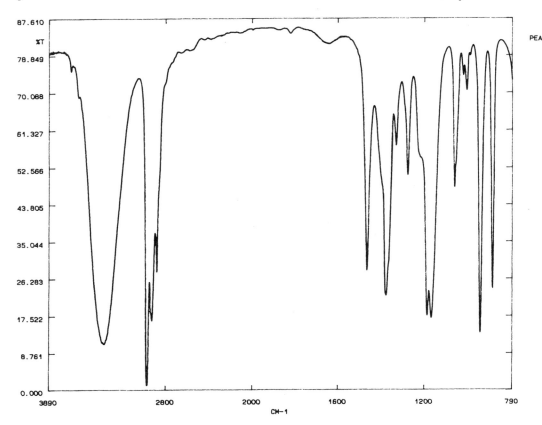

PEAK THRESHOLD = 0.87 %T
     CM-1   %T

     3682.00   75.375
     3367.00   11.215
     2971.00   1.285
     2927.00   16.640
     2883.00   28.145
     2360.00   74.780
     2342.00   77.010
     1820.00   84.285
     1646.00   78.275
     1465.00   28.375
     1376.00   22.375
     1331.00   57.860
     1277.00   50.920
     1186.00   17.720
     1167.00   17.155
     1060.00   48.015
     1020.00   74.275
     1004.00   70.800
      940.00   13.685
      882.00   24.350

Sample 1.1

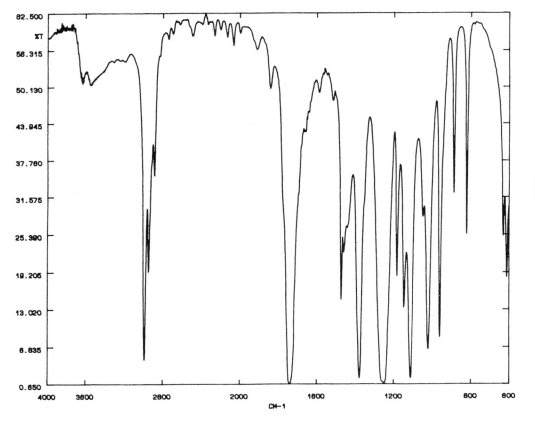

     CM-1   %T

     3454.00   46.652
     2983.00   4.745
     2942.00   19.302
     2883.00   35.202
     2087.00   57.147
     1841.00   49.882
     1741.00   0.852
     1471.00   14.780
     1375.00   1.692
     1246.00   0.755
     1183.00   18.652
     1146.00   13.407
     1111.00   1.730
     1020.00   8.520
      960.00   8.477
      868.00   32.290
      821.00   25.465
      829.00   25.197
      811.00   18.382

Sample 1.2

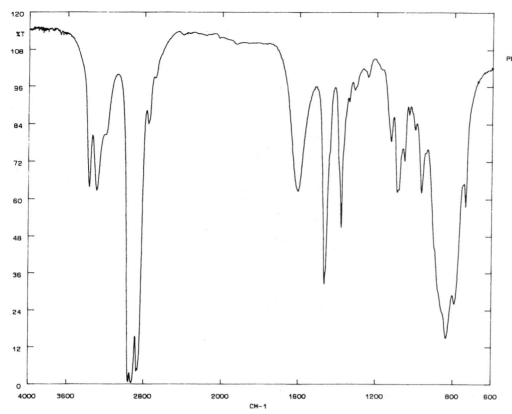

PEAK THRESHOLD = 12.00 %T	
CM-1	%T
3367.00	63.705
3290.00	62.580
2924.00	0.300
1605.00	62.450
1465.00	32.335
1378.00	50.800
1090.00	62.150
967.00	62.030
837.00	14.860

Sample 1.3

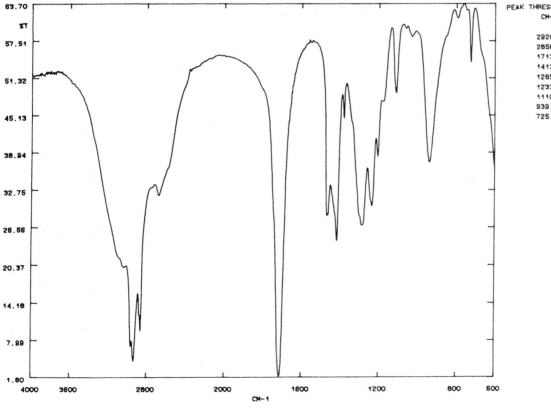

PEAK THRESHOLD = 6.19 %T	
CM-1	%T
2928.00	4.512
2858.00	9.542
1713.00	1.802
1413.00	24.160
1285.00	26.767
1233.00	29.965
1110.00	48.802
939.00	37.175
725.00	53.837

Sample 1.4

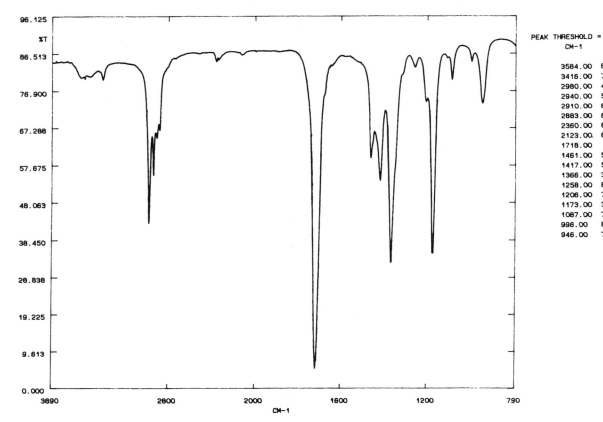

PEAK THRESHOLD = 0.96 %T
       CM-1   %T

     3584.00   80.210
     3416.00   79.970
     2980.00   43.075
     2940.00   55.345
     2910.00   64.900
     2883.00   66.895
     2360.00   84.670
     2123.00   86.415
     1718.00   5.410
     1461.00   59.675
     1417.00   53.885
     1366.00   32.830
     1258.00   83.160
     1206.00   74.245
     1173.00   35.205
     1087.00   79.970
      996.00   84.585
      946.00   73.845

Sample 1.5

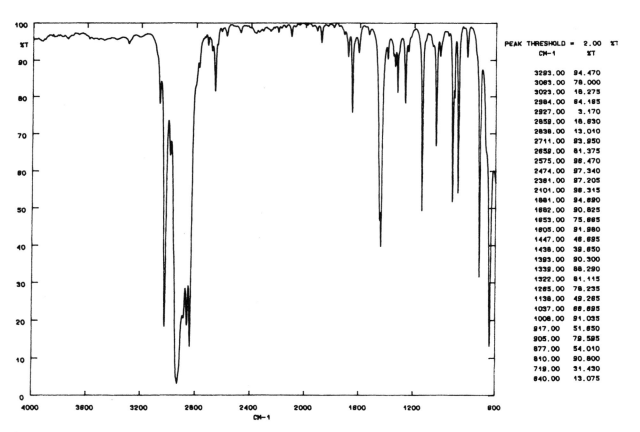

PEAK THRESHOLD = 2.00   %T
       CM-1   %T

     3293.00   94.470
     3063.00   78.000
     3023.00   18.275
     2984.00   84.185
     2927.00   3.170
     2859.00   18.630
     2838.00   13.010
     2711.00   93.950
     2659.00   81.375
     2575.00   96.470
     2474.00   97.340
     2361.00   97.205
     2101.00   96.315
     1881.00   94.890
     1662.00   90.825
     1653.00   75.665
     1605.00   91.980
     1447.00   48.695
     1438.00   39.850
     1393.00   90.300
     1339.00   88.290
     1322.00   81.115
     1265.00   78.235
     1138.00   49.285
     1037.00   88.695
     1008.00   91.035
      917.00   51.850
      905.00   79.595
      877.00   54.010
      810.00   90.800
      719.00   31.430
      640.00   13.075

Sample 1.6

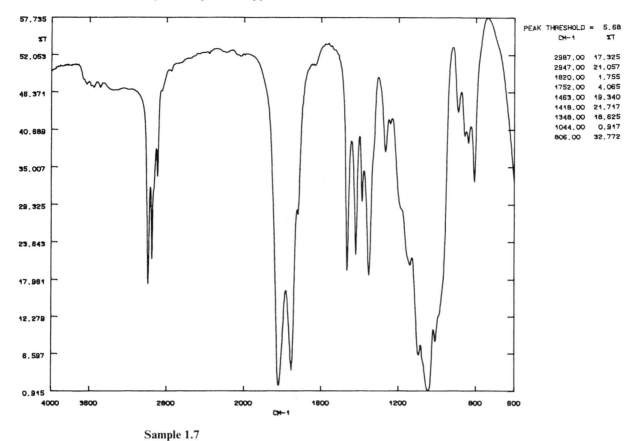

Sample 1.7

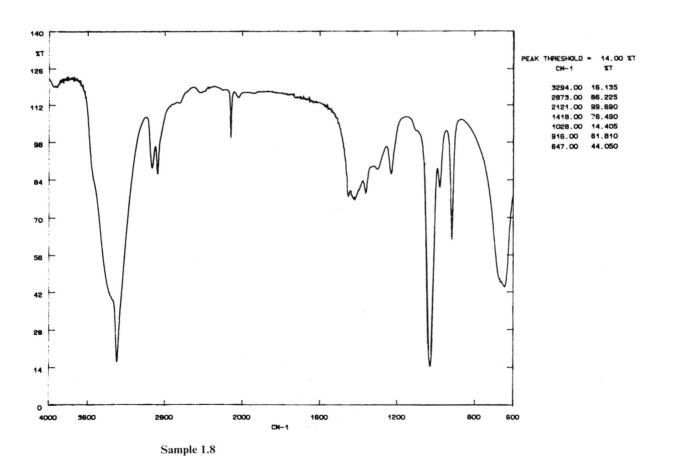

Sample 1.8

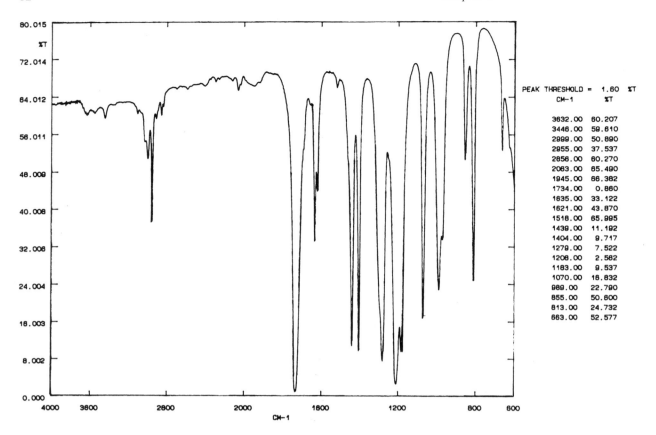

PEAK THRESHOLD =   1.60 %T
      CM-1   %T

    3632.00   60.207
    3446.00   59.610
    2999.00   50.890
    2955.00   37.537
    2856.00   60.270
    2063.00   65.490
    1945.00   66.382
    1734.00   0.860
    1635.00   33.122
    1621.00   43.870
    1518.00   65.995
    1439.00   11.192
    1404.00   9.717
    1279.00   7.522
    1208.00   2.582
    1183.00   9.537
    1070.00   16.832
     989.00   22.790
     855.00   50.800
     813.00   24.732
     663.00   52.577

Sample 1.9

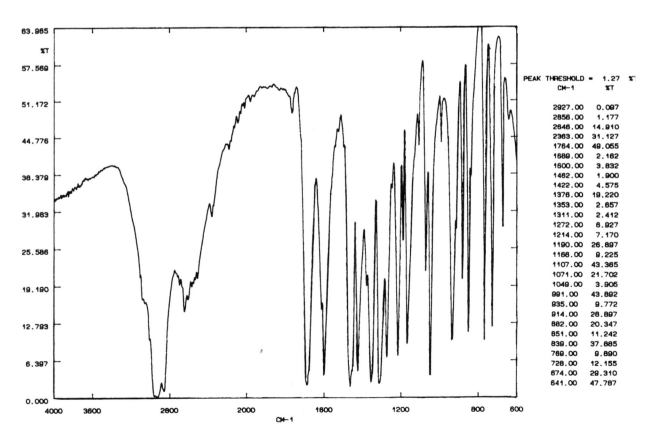

PEAK THRESHOLD =   1.27 %T
      CM-1   %T

    2927.00   0.097
    2856.00   1.177
    2646.00   14.910
    2363.00   31.127
    1764.00   49.055
    1689.00   2.162
    1600.00   3.832
    1462.00   1.900
    1422.00   4.575
    1376.00   19.220
    1353.00   2.657
    1311.00   2.412
    1272.00   6.927
    1214.00   7.170
    1190.00   26.897
    1166.00   9.225
    1107.00   43.385
    1071.00   21.702
    1049.00   3.905
     991.00   43.892
     935.00   9.772
     914.00   28.897
     882.00   20.347
     851.00   11.242
     839.00   37.885
     769.00   9.890
     728.00   12.155
     674.00   29.310
     641.00   47.787

Sample 1.10

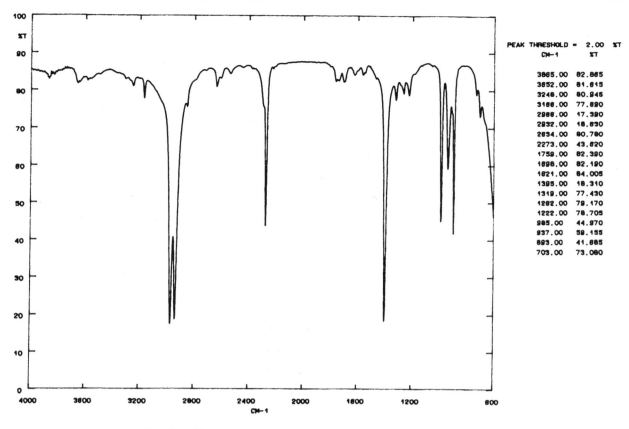

PEAK THRESHOLD = 2.00 %T
CM-1   %T

3865.00   82.865
3652.00   81.615
3248.00   80.945
3186.00   77.690
2966.00   17.390
2932.00   18.630
2634.00   80.780
2273.00   43.620
1759.00   82.390
1698.00   82.190
1621.00   84.005
1395.00   18.310
1319.00   77.430
1282.00   79.170
1222.00   78.705
985.00   44.970
937.00   59.155
893.00   41.685
703.00   73.080

Sample 1.11

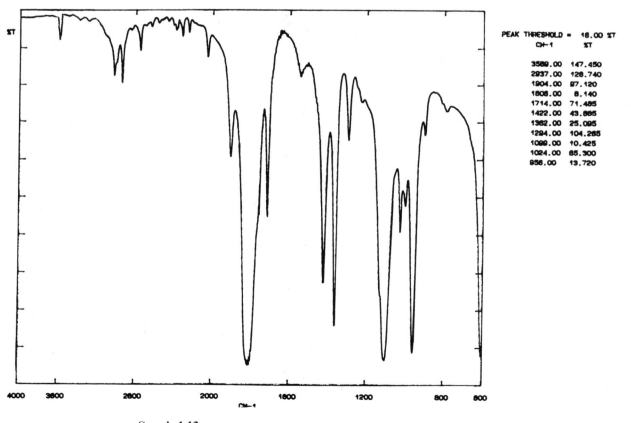

PEAK THRESHOLD = 16.00 %T
CM-1   %T

3589.00   147.450
2937.00   128.740
1904.00   97.120
1608.00   8.140
1714.00   71.485
1422.00   43.885
1362.00   25.095
1294.00   104.265
1099.00   10.425
1024.00   65.300
956.00   13.720

Sample 1.12

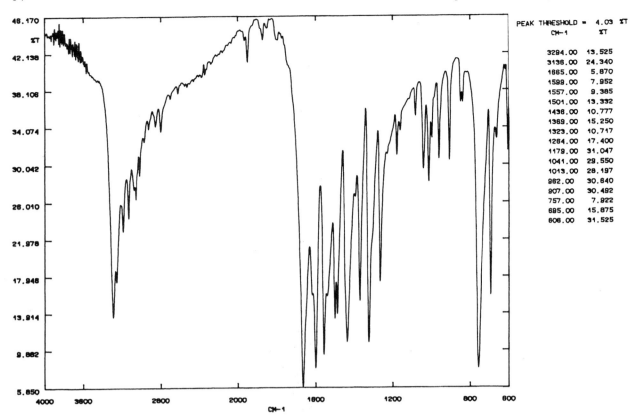

PEAK THRESHOLD = 4.03 %T
     CM-1   %T

    3294.00   13.525
    3136.00   24.340
    1665.00   5.870
    1599.00   7.952
    1557.00   9.385
    1501.00   13.332
    1436.00   10.777
    1369.00   15.250
    1323.00   10.717
    1264.00   17.400
    1179.00   31.047
    1041.00   29.550
    1013.00   28.197
     982.00   30.640
     907.00   30.492
     757.00   7.922
     695.00   15.875
     606.00   31.525

Sample 1.13

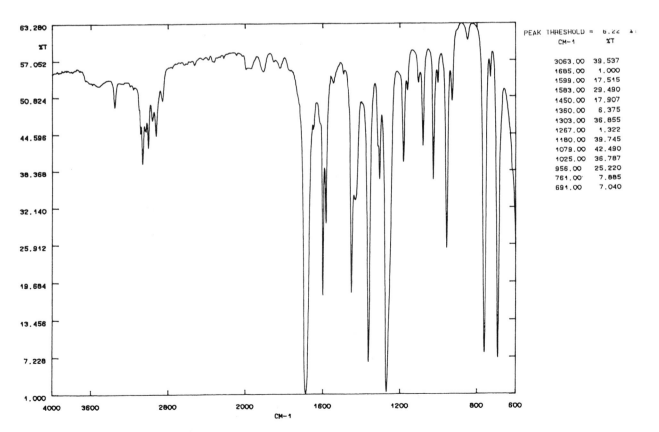

PEAK THRESHOLD = 6.22 %T
     CM-1   %T

    3063.00   39.537
    1685.00   1.000
    1599.00   17.515
    1583.00   29.490
    1450.00   17.907
    1360.00   6.375
    1303.00   36.855
    1267.00   1.322
    1180.00   39.745
    1079.00   42.490
    1025.00   36.787
     956.00   25.220
     761.00   7.885
     691.00   7.040

Sample 1.14

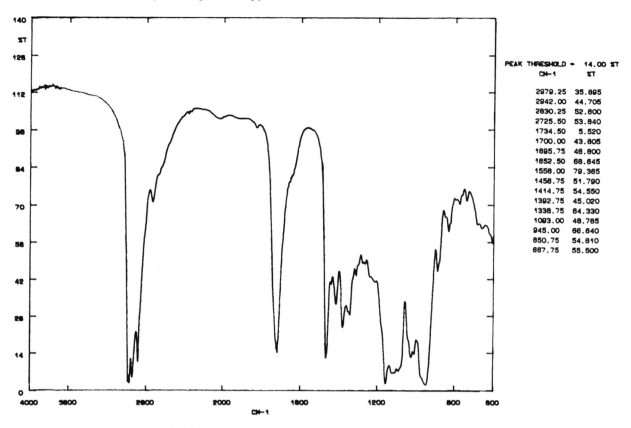

PEAK THRESHOLD = 14.00 %T

CM-1	%T
2979.25	35.895
2942.00	44.705
2830.25	52.800
2725.50	53.840
1734.50	5.520
1700.00	43.805
1695.75	46.800
1652.50	68.645
1558.00	79.365
1456.75	51.790
1414.75	54.550
1392.75	45.020
1338.75	64.330
1093.00	48.785
945.00	66.640
850.75	54.810
667.75	55.500

Sample 1.15

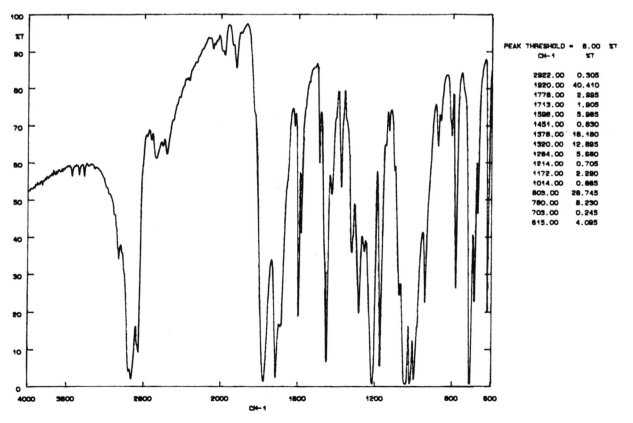

PEAK THRESHOLD = 8.00 %T

CM-1	%T
2922.00	0.305
1920.00	40.410
1778.00	2.995
1743.00	1.905
1598.00	5.985
1451.00	0.830
1378.00	18.180
1320.00	12.895
1264.00	5.680
1214.00	0.705
1172.00	2.290
1014.00	0.885
803.00	26.745
780.00	8.230
703.00	0.245
615.00	4.085

Sample 1.16

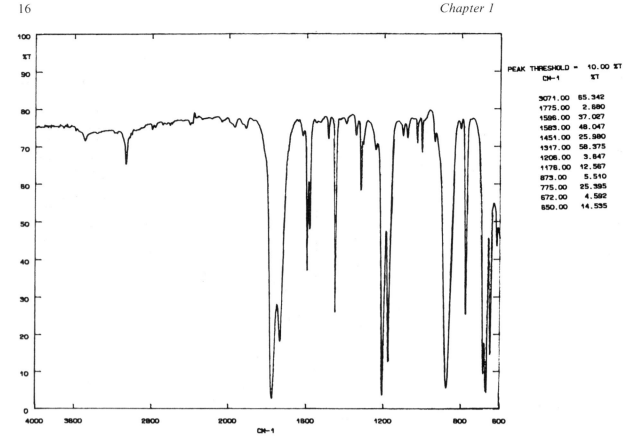

PEAK THRESHOLD = 10.00 %T

CM-1	%T
3071.00	85.342
1775.00	2.680
1596.00	37.027
1583.00	48.047
1451.00	25.980
1317.00	58.375
1206.00	3.847
1176.00	12.567
873.00	5.510
775.00	25.395
672.00	4.592
650.00	14.535

Sample 1.17

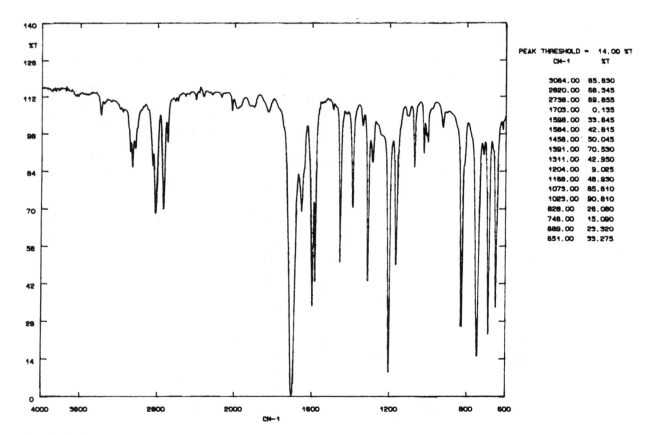

PEAK THRESHOLD = 14.00 %T

CM-1	%T
3064.00	85.830
2820.00	68.345
2738.00	89.855
1703.00	0.135
1598.00	33.845
1584.00	42.815
1456.00	50.045
1391.00	70.530
1311.00	42.950
1204.00	9.025
1166.00	48.930
1073.00	85.810
1023.00	90.810
826.00	26.080
746.00	15.090
689.00	23.320
651.00	33.275

Sample 1.18

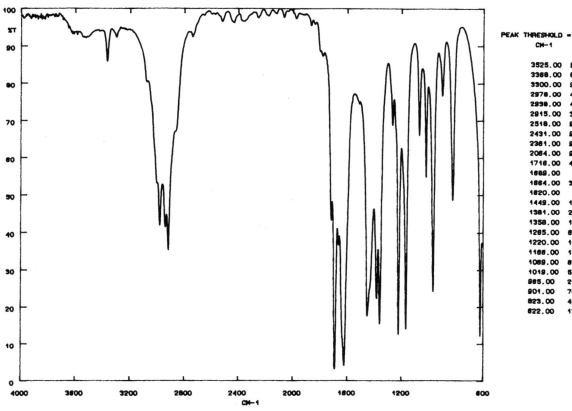

PEAK THRESHOLD = 2.00 %T

CM-1	%T
3525.00	92.175
3366.00	85.780
3300.00	92.320
2978.00	41.710
2938.00	41.330
2915.00	35.255
2518.00	96.450
2431.00	96.140
2361.00	91.640
2064.00	97.590
1718.00	43.175
1689.00	3.170
1664.00	36.630
1620.00	4.105
1449.00	17.560
1381.00	22.255
1358.00	15.440
1265.00	68.755
1220.00	12.665
1166.00	14.010
1069.00	86.035
1019.00	54.840
965.00	24.170
901.00	76.595
823.00	46.845
622.00	12.175

Sample 1.19

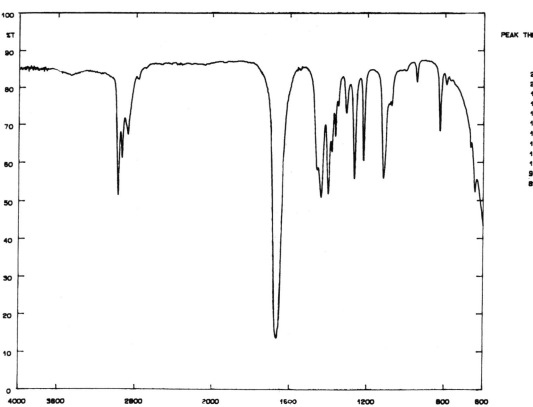

PEAK THRESHOLD = 5.00 %T

CM-1	%T
2978.00	51.495
2939.00	81.310
1666.00	13.610
1436.00	50.780
1401.00	51.715
1385.00	68.835
1309.00	72.995
1265.00	55.725
1218.00	80.380
1114.00	55.830
944.00	81.405
824.00	68.410

Sample 1.20

CHAPTER 2

Mass Spectrometry

Mass spectrometry gives us some information about the structure of a molecule. It can usually tell us the molecular weight of a substance, and can detect the presence of bromine, chlorine and iodine. It may be possible to deduce the structure of a compound from its mass spectrum.

Unlike the other techniques studied in this book, it is a spectrometric, not a spectroscopic, method. It does not involve the absorption of radiation; it consists of generating ions of such energy that they fragment, collecting fragments of the same mass together, and weighing them. Some fragments are more stable than others, and so are produced to a much greater extend than others, and we see large peaks in the spectrum at their mass. Using knowledge of favoured fragmentations, a surprising amount of information about the structure of the original molecule can be obtained.

Mass spectrometers operate at very low pressures, so that the ions can travel long distances without undergoing collisions. The mass spectrometer consists basically of three parts, as shown schematically in Diagram 2.1. The regions of the spectrometer can be described according to their function, as ionisation, separation and collection.

Ionisation

The material to be studied is introduced into the ionisation chamber in the vapour phase. It is bombarded with a stream of electrons, generated by the filament and accelerated to about 70 eV by the anode.

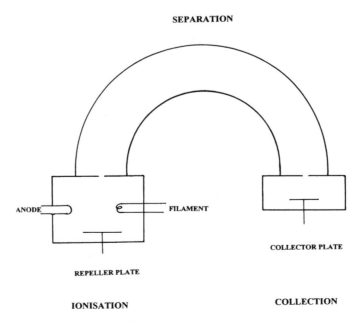

Diagram 2.1 *Schematic diagram of a mass spectrometer*

If an electron at this energy hits one of the atoms being studied, the atom gains considerable vibrational energy, causing it to ionise and usually fragment. The ions produced pass through a series of slits into the analysing system.

This method of ionisation is used for most samples which are reasonably stable and reasonably volatile. A number of other ionisation methods exist to tackle less stable or less volatile samples, but all work on the same principle: the molecule absorbs energy and breaks down to its more stable ions.

Ion separation

Ionisation of a compound gives a series of ions of different masses, but almost all carry a single positive charge. The stream of positive ions needs to be separated according to the mass to charge ratio, usually denoted by m/z, which effectively separates them by mass, as $z = 1$. This is done by passing the beam through a magnetic field, which deflects lighter ions more than heavier ions. The actual amount of deflection is dependent on the mass to charge ratio, m/z. At the end of the travel through the magnetic field, the beam passes through a narrow slit, and proceeds to the collector. By varying the magnetic field, ions of any m/z ratio can be focused on the collector slit, and the spectrum scanned.

Most mass spectrometers use a more complex arrangement of electrostatic and magnetic analysers in order to improve the resolution of the instrument, but the principle involved is the same.

Ion collection

When the ions reach the collector, ions of each m/z ratio land on the collector plate for a similar interval of time, and build up a charge which is amplified and recorded.

One of the great advantages of mass spectrometry is that it can work with extremely small samples, less than 10^{-12} of a mole. It is often used coupled to a gas–liquid chromatography analyser, and the separated fractions are run automatically into the mass spectrometer. It is particularly valuable for the study of biological samples.

The mass spectrometer ionises the molecules, then divides the ions up according to their m/z ratio, and measures the amount of ions of each m/z ratio. The result is presented as a bar chart, as shown in Figure 2.1, the mass spectrum of *n*-octane, $CH_3CH_2CH_2CH_2CH_2CH_2CH_2CH_3$.

Along the base of the chart, the scale is in m/z ratio values, but since $z = 1$, these can be regarded as the masses of the ions. The vertical scale shows the amount of each ion. Since the amount of any ion collected obviously depends on how long the collector plate has been in the ion beam, we adopt an arbitrary scale for intensity, in which the largest peak is given an intensity value of 100 and is known as the parent peak. All other peaks have their intensities measured as a percentage of the parent peak.

Before we start to look at the structural information which can be obtained, we need to consider the information which can be obtained from isotope masses and ratios. One of the most useful things we can do is to measure the accurate mass of a peak. The atomic weights of most isotopes are not quite integral numbers, and most have been measured accurately to six places of decimals. A few useful accurate masses are given in Table 2.1.

If we measure the molecular weight of any ion to six-figure accuracy, on the basis of that molecular weight we can assign a unique chemical composition to

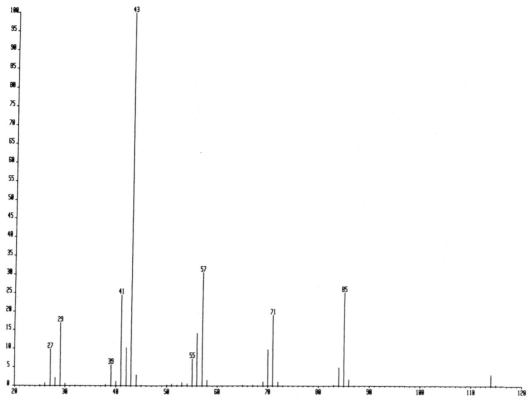

Figure 2.1

Table 2.1 *Accurate atomic weights of some common isotopes*

Isotope	Atomic weight
^{1}H	1.007825
^{12}C	12.000000
^{14}N	14.003074
^{16}O	15.994915
^{35}Cl	34.968855
^{79}Br	78.918348
^{127}I	126.904352

that ion. As an example, assume we have an ion of m/z ratio 28. The composition of this ion could be N_2, C_2H_4 or CO. If we calculate the exact weights of these ions, we find they differ significantly:

$^{14}N_2$ has molecular weight 28.006148
$^{12}C_2{}^1H_4$ has molecular weight 28.031300
$^{12}C^{16}O$ has molecular weight 27.994915

Thus, we can identify the peak of m/z 28 by measuring its accurate mass. We do not have to worry about the isotopic composition of C, N or H, as the presence of ^{2}H, ^{15}N or ^{13}C would result in a separate peak of m/z greater than 28.

The method requires a larger sample than can be provided by gas chromatography–mass spectrometry (GC/MS), and is time consuming, but it remains a valuable approach to the determination of molecular composition.

We saw that in order to define the composition of the ion, we had to specify its isotopic as well as its chemical composition. Some elements exist as

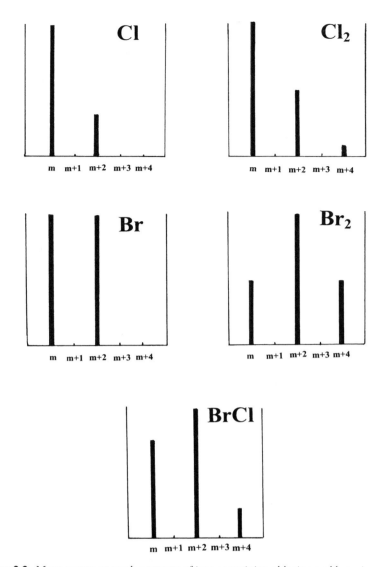

Diagram 2.2 *Mass spectrum peak patterns of ions containing chlorine and bromine*

mixtures of isotopes in which appreciable amounts of each isotope are present, and then we always get a mixture of ions of different isotopic composition. Since chlorine has two isotopes, ^{35}Cl with an abundance of 75.8% and ^{37}Cl with an abundance of 24.2%, then an ion containing chlorine will always have two peaks, separated by two mass units, in the ratio of approximately 3 to 1. Similarly, bromine consists of 50.5% of ^{79}Br and 49.5% of ^{81}Br, so ions containing bromine always consist of a pair of peaks of equal size, separated by two mass units. The patterns produced, together with those resulting from Cl_2, Br_2 and BrCl, are shown in Diagram 2.2. Detection of these patterns is valuable evidence for the presence of chlorine or bromine in a molecule, especially as these elements are difficult to detect by infrared spectroscopy. Iodine exists as a single isotope, but has an atomic weight of 127, so it shows up very clearly in a mass spectrum.

Most other isotope compositions favour one isotopic species, but even these can give rise to significant extra peaks. Carbon is a mixture of 98.9% ^{12}C and 1.1% ^{13}C, so that a compound containing one carbon atom per molecule would have peaks in the ratio 98.9 to 1.1. However, if an ion has ten carbon atoms, we have ten chances of a ^{13}C atom occurring in the ion, and the ion would have two peaks in the ratio of 89 to 11. Any sizeable peak in a mass spectrum of an organic compound is thus likely to be followed by a peak 1 mass unit larger, whose size depends on the number of carbon atoms in the ion.

We can now return to the spectrum of *n*-octane, shown in Figure 2.1. The first thing to look for when we study a mass spectrum is the molecular ion, which gives us the molecular weight. In this case, it is the rather small peak at m/z 114. One of the problems of mass spectrometry is identifying the molecular ion; in molecules containing only C, H, O and halogens, it is useful to note that the main peaks have odd m/z ratios, while the molecular ion has an even m/z ratio. If the molecule contains an odd number of nitrogen atoms this is reversed: the molecular ion has an odd m/z ratio, and most of the peaks have even m/z ratios. Having identified the molecular ion, we look at the differences in m/z ratios between the molecular ion and the main peaks in the spectrum, which gives us the number of mass units (known as Daltons) lost in each fragmentation. In the case of octane, these are:

Molecular ion		Ion		Mass lost	Possible fragment
114	–	85	=	29	C_2H_5
114	–	71	=	43	C_3H_7
114	–	57	=	57	C_4H_9
114	–	43	=	71	C_5H_{11}

We can then look at the most probable compositions of the fragments lost. In this case, we know that the molecule is octane, so can ignore possible compositions which contain oxygen. Even if we did not know this, we could say that the fragment of mass 43 Daltons which was lost cannot be CH_3CO, since that structural unit would also split next to the C–O linkage (see page 26), losing CH_3 with a mass of 15 Daltons. A list of possible compositions of fragment ions is given in Table 2.2.

Our octane molecule thus has three main fragmentations:

$$CH_3\text{–}CH_2 \mid CH_2 \mid CH_2 \mid CH_2\text{–}CH_2\text{–}CH_2\text{–}CH_3$$

Knowing that the molecule can split off C_2H_5, C_3H_7 or C_4H_9 and has a molecular weight of 114 provides good evidence for the structure of *n*-octane; loss of fragments which increase in mass by 14 (*i.e.* CH_2) is characteristic of an unbranched hydrocarbon chain. Note that octane does not readily lose CH_3. The fission of this species is energetically unfavourable, unless the residual ion is stabilised by an electron rich atom or group.

Table 2.2 *Possible losses from molecular ions*

Mass lost	Possible group lost	Inference
15	CH_3	
17	OH	Alcohol
18	H_2O	Alcohol
29	C_2H_5 or CHO	–
31	OCH_3	Methyl ester
41	C_3H_5	Propyl ester
43	C_3H_7 or CH_3CO	–
44	CO_2	Anhydride
45	COOH	Carboxylic acid
	OC_2H_5	Ethyl ester
55	C_4H_7	Butyl ester
57	C_4H_9 or C_2H_5CO	–
71	C_5H_{11} or C_3H_7CO	–

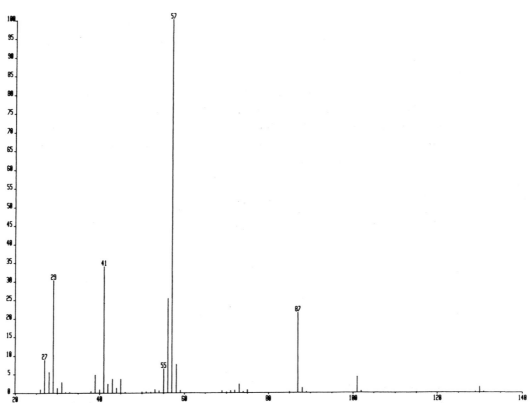

Figure 2.2

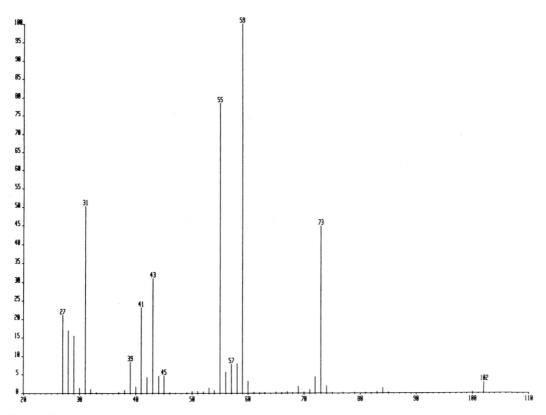

Figure 2.3

As soon as we introduce a different atom into our molecule, the simple picture we have seen for octane changes. Figure 2.2 shows the mass spectrum of di-*n*-butyl ether, $CH_3CH_2CH_2CH_2OCH_2CH_2CH_2CH_3$. We have a molecular ion at 130, but only three fragmentations, one of which dominates. These are:

Molecular ion		Ion		Mass lost	Possible fragment
130	–	101	=	29	C_2H_5
130	–	87	=	43	C_3H_7
130	–	57	=	73	C_4H_9O

In this case, the oxygen atom affects the pattern of splitting of the molecule by stabilising the ion containing oxygen; the main splitting is that of the C–O bond. Loss of C_3H_7 also occurs because the ion $CH_2=O^+C_4H_9$ is also strongly stabilised by the oxygen (a nitrogen heteroatom gives a similar stabilisation and splitting):

$$CH_3\text{–}CH_2 \mid CH_2 \mid CH_2 \mid O\text{–}CH_2\text{–}CH_2\text{–}CH_2\text{–}CH_3$$

The next spectrum returns to one of the persistent problems of mass spectrometry, which is to identify the molecular ion. This is particularly difficult with the spectra of alcohols, where the molecular ion is often small or absent, as is shown in Figure 2.3, the spectrum of hexan-3-ol, $CH_3CH_2CH(OH)CH_2CH_2CH_3$.

The small peak at *m/z* 102 is the molecular ion. The main fragmentations of the ion are:

Molecular ion		Ion		Mass lost	Possible fragment
102	–	73	=	29	C_2H_5
102	–	59	=	43	C_3H_7
102	–	55	=	47	
102	–	43	=	59	C_3H_9O

The main splittings are thus:

$$\begin{array}{c} OH \\ | \\ CH_3\text{–}CH_2 \mid CH \mid CH_2\text{–}CH_2\text{–}CH_3 \end{array}$$

The –OH group stabilises an ion generated at the carbon to which it is attached, so splitting of either of the C–C bonds to that carbon atom is favoured.

Note that we have not considered the peak at *m/z* 55, which has resulted from splitting off a mass of 47. There is not a reasonable fragment of mass 47 which can split off hexan-3-ol. The peak at *m/z* 55 probably arises after an initial dehydration of hexan-3-ol to hexene. Loss of water (mass 18 Daltons) would give hexene, molecular weight 84. The spectrum in fact shows a small peak at *m/z* 84. If hexene loses C_2H_5 (mass 29 Daltons), then we would expect a peak at *m/z* 55.

If we had made an error in our choice of the molecular ion peak, then the observed fragmentations would have been impossible. Suppose we thought the molecular ion was at *m/z* 84, which is possible since it involves an even mass. The peak at *m/z* 73 would then involve loss of 11 Daltons, and that at 59

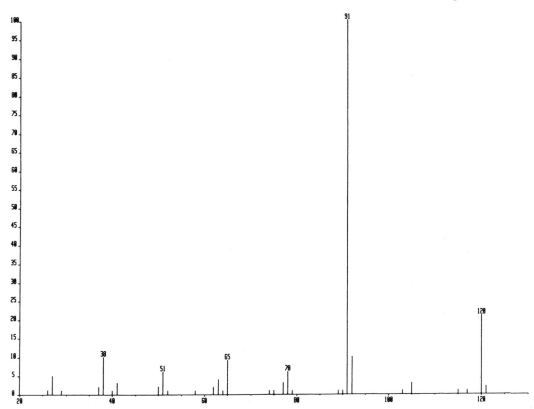

Figure 2.4

the loss of 25 Daltons. Neither fragmentation involves a reasonable ion, so, clearly, something is wrong.

If we have the mass spectrum of a molecule containing an aromatic ring, then we should expect to find a peak from a very stable ion, the benzyl ion, $PhCH_2^+$, provided the structure permitted this ion to form. An unsubstituted benzyl peak is found at m/z 91; substitution will obviously change the mass of the ion. Formation of a benzyl ion is shown in the next spectrum, Figure 2.4, which shows the mass spectrum of propylbenzene, $PhCH_2CH_2CH_3$.

The spectrum shows a large molecular ion peak at m/z 120, and a single large peak at m/z 91, resulting from the loss of 29 Daltons, consistent with the loss of C_2H_5. The peak at m/z 91 is the benzyl ion, and the molecule shows a single fragmentation:

$$PhCH_2 \mid CH_2\text{--}CH_3$$

If an aromatic molecule cannot form a benzyl ion, then the phenyl ion $C_6H_5^+$, m/z 77, may be formed, but formation of the benzyl ion is favoured.

The next spectrum is that of octan-3-one, $CH_3CH_2COCH_2CH_2CH_2CH_2CH_3$, shown in Figure 2.5.

On the basis of our analysis of the spectrum of hexan-3-ol, we would expect this molecule to split on each side of the carbonyl carbon atom, since this would yield a stabilised acylium ion, $R\text{--}C{=}O^+$. The spectrum shows that this is in fact observed:

Molecular ion		Ion		Mass lost	Possible fragment
128	–	99	=	29	C_2H_5
128	–	72	=	56	
128	–	57	=	71	C_5H_{11}

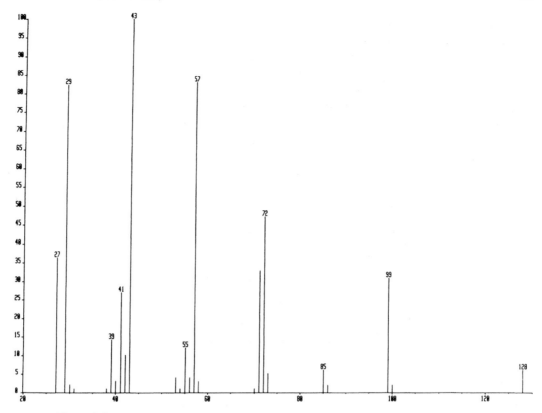

Figure 2.5

These splittings yield peaks at m/z 99 and 57. There is, however, another peak at m/z 72. It has an *even* m/z value; all major peaks which we have observed previously have odd m/z values. It is a general guideline that the mass spectrum of an organic compound containing no element other than carbon, hydrogen, oxygen and the halogens has an even m/z molecular ion and odd m/z fragment ions.

So far, we have noted two variations on this rule:

A. The molecule contains an odd number of nitrogen atoms. In this case, the molecular ion will have an odd m/z ratio, and fragment ions containing an odd number of nitrogen atoms an even m/z ratio.

B. The molecule has undergone an elimination reaction to give a stable product. We have already seen that hexan-3-ol, which has a molecular ion at m/z 102, can lose water to give hexene, which has a molecular ion at m/z 84. Carboxylic acids often lose CO_2 (mass 44 Daltons) to give an ion of even m/z ratio; loss of HCl (masses 36 and 38 Daltons) and HBr (masses 80 and 82 Daltons) and loss of ethene (mass 28 Daltons) are also observed. Such peaks are often small.

This observation now adds a third reason for a peak to have an even m/z ratio, namely the McLafferty rearrangement:

The McLafferty rearrangement occurs whenever we have a carbonyl compound which has a γ-hydrogen atom in a position which permits it to migrate to the carbonyl oxygen atom through a cyclic transition state. The reaction can occur with any type of carbonyl group, and the *m/z* ratio of the even peak tells us what other substituents there are on the carbonyl carbon atom (Table 2.3). If we have an amide, so that X = NH_2, then the McLafferty rearrangement gives a peak of odd *m/z* ratio, 59. If the carbonyl group has a substituent on the neighbouring carbon atom, then a McLafferty rearrangement gives a different size of even *m/z* peak.

Table 2.3 *Ions characteristic of McLafferty rearrangements*

Compound	X	Ion (m/z)
Aldehyde	H	44
Methyl ketone	CH_3	58
Carboxylic acid	OH	60
Ethyl ketone	CH_3	72
Methyl ester	OCH_3	74
Ethyl ester	OC_2H_5	88

Summary

When we first look at a mass spectrum, we should look for:

A. The molecular ion. This often, but not always, has an even *m/z* ratio, and it may be very small.

B. The presence of the halogens. Bromine and chlorine are readily spotted by peak patterns; bromine has two equal peaks two mass units apart, while chlorine has two peaks in the ratio 3 to 1 which are also two mass units apart. Iodine can be spotted from sheer size: it has an atomic weight of 127.

C. Any major ion of even *m/z* ratio. If the molecular ion has an odd *m/z* ratio, then the ion contains an odd number of nitrogen atoms; if it has an even *m/z* number, then you probably have an elimination reaction or a McLafferty rearrangement of a carbonyl compound. Look also for ions at *m/z* 91 (possibly benzyl) and *m/z* 77 (possibly phenyl).

D. The main ions in the spectrum, and calculate what size of unit has been split from the molecular ion. Look at Table 2.2 to see the possible identity of the fragment split off. From this information you may be able to determine the original structure of the sample.

Problems in Interpreting Mass Spectra

You are given the mass spectra of 20 samples. Try to identify the structures of the samples from the mass spectra.

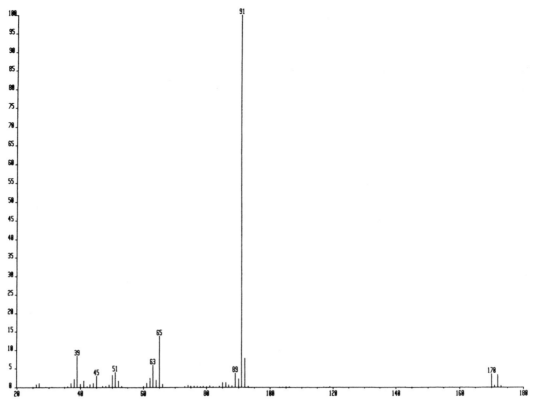

Sample 2.1

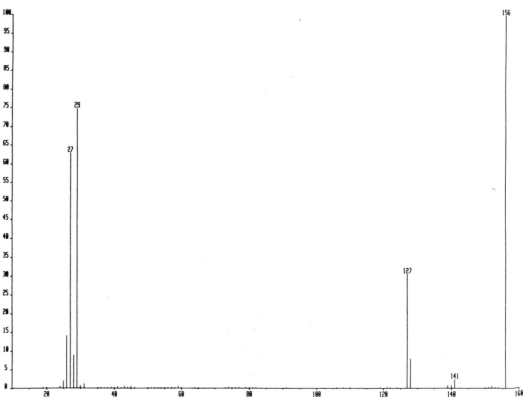

Sample 2.2

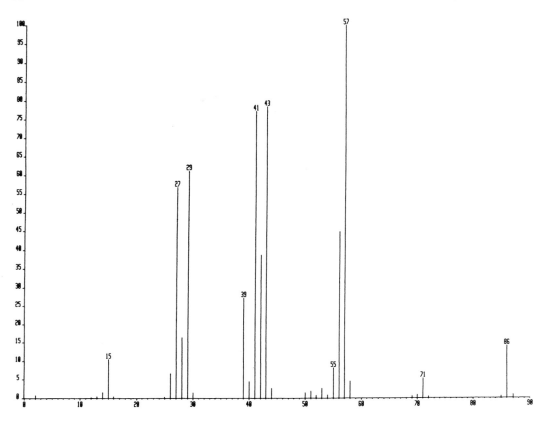

Sample 2.3

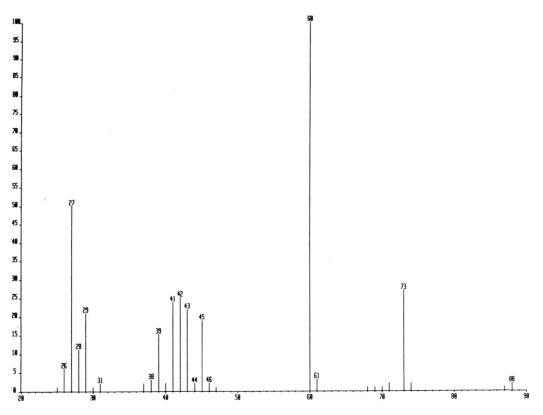

Sample 2.4

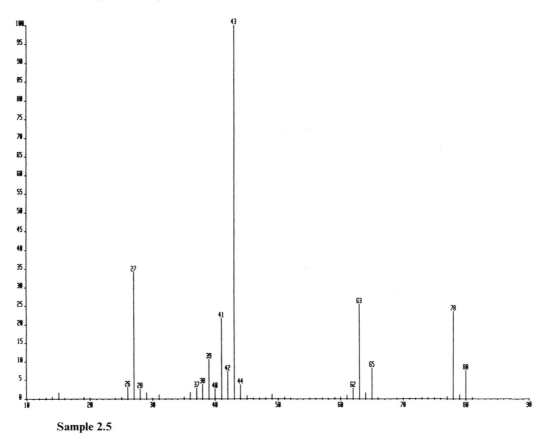

Sample 2.5

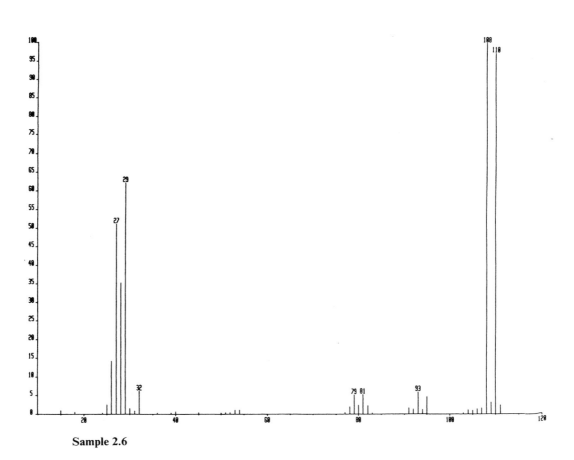

Sample 2.6

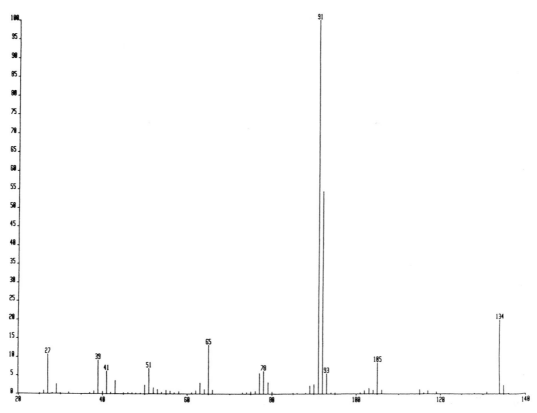

Sample 2.7

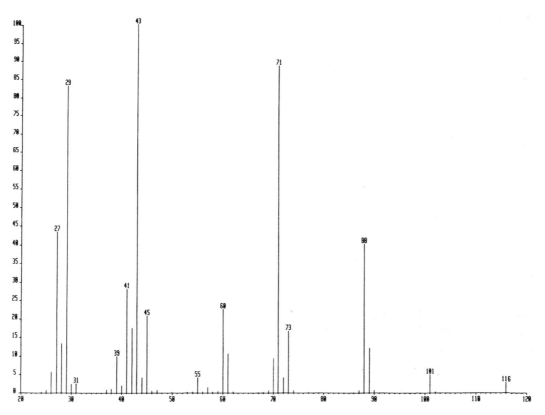

Sample 2.8

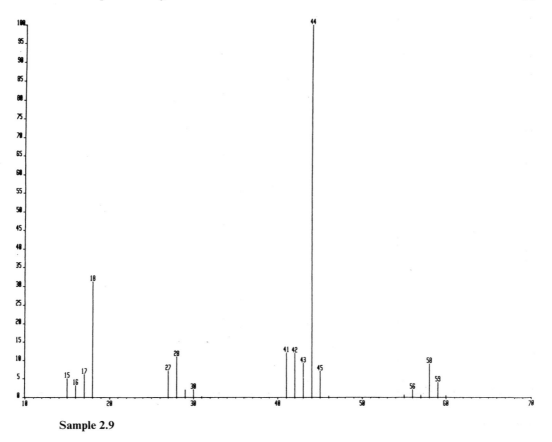

Sample 2.9

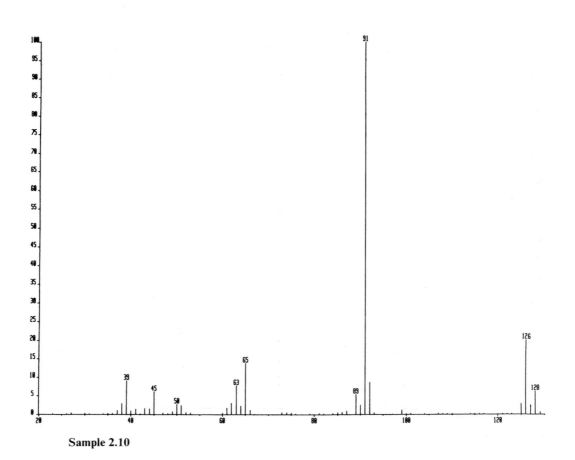

Sample 2.10

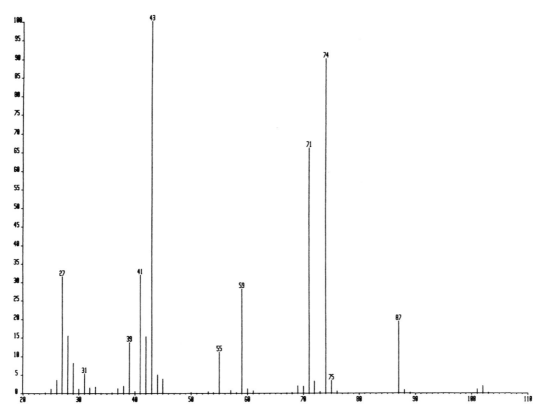

Sample 2.11

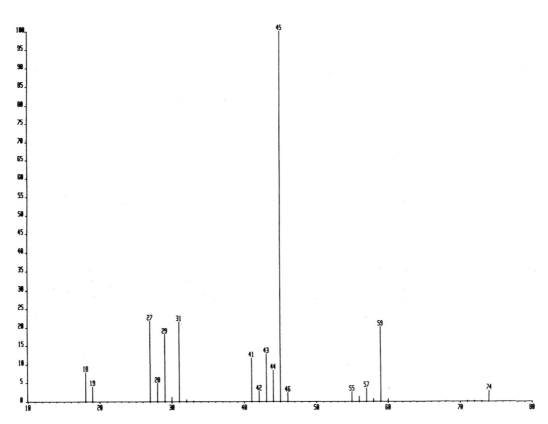

Sample 2.12

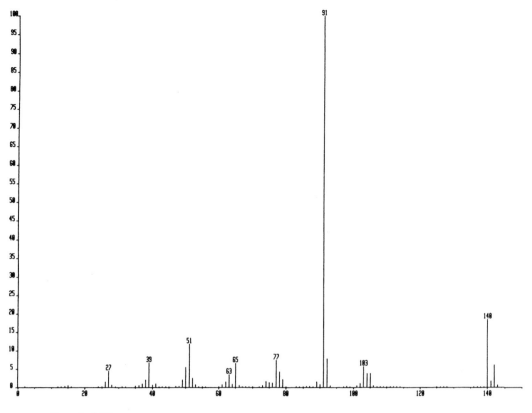

Sample 2.13

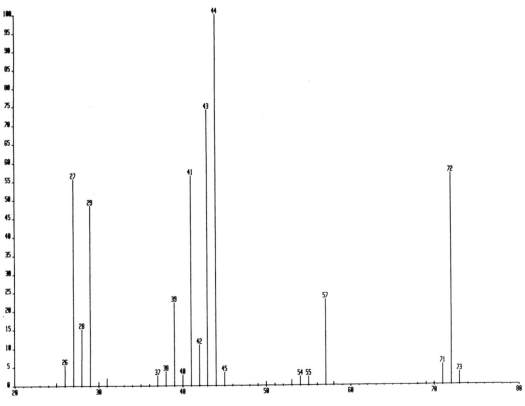

Sample 2.14

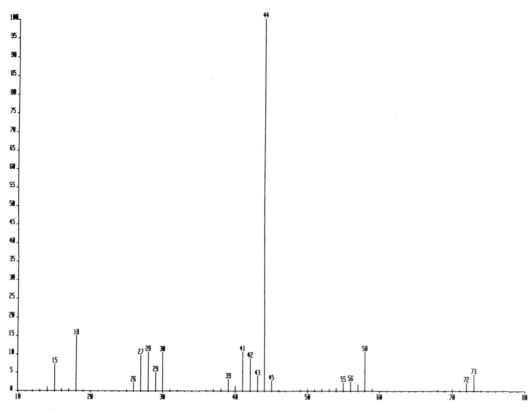

Sample 2.15

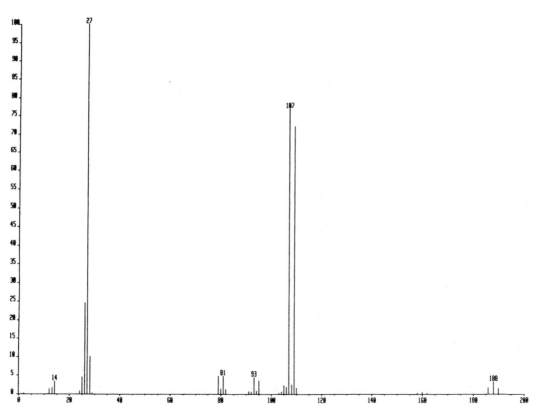

Sample 2.16

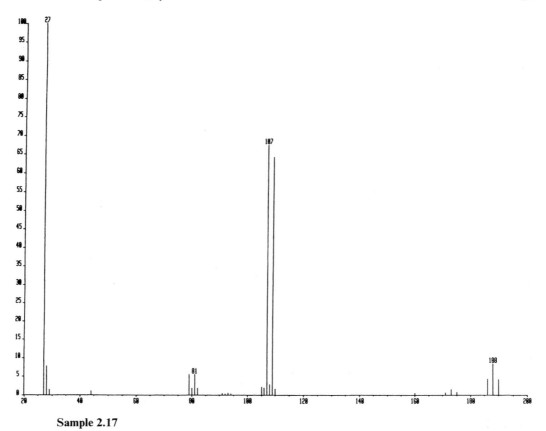

Sample 2.17

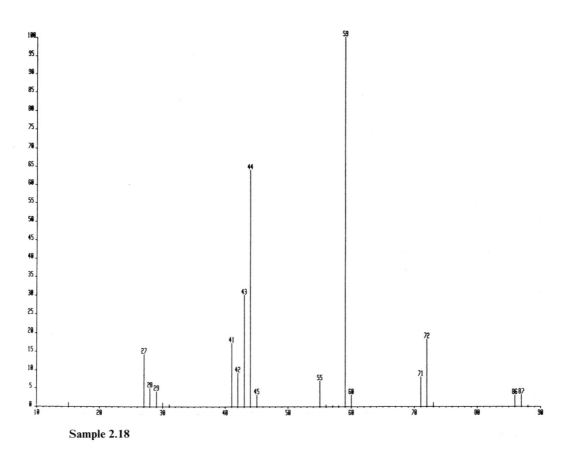

Sample 2.18

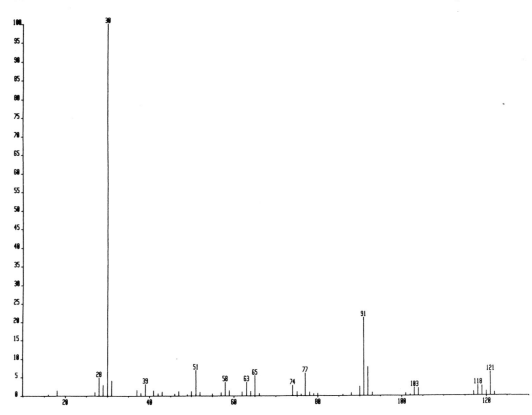

Sample 2.19

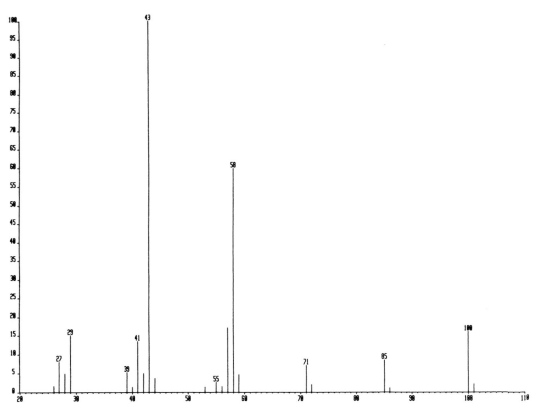

Sample 2.20

CHAPTER 3

Problems in Interpreting Infrared Spectra and Mass Spectra

We have now seen how to identify functional groups by infrared spectroscopy and to find the molecular weight, halogen content and structural features of a molecule by mass spectrometry. If we use both techniques on a sample, then we can substantially extend the range of molecules we can identify. Both infrared spectroscopy and mass spectrometry can identify almost any molecule by comparison with a spectrum of the authentic material, but when working entirely with unknowns it is often difficult to distinguish isomers. The substances whose spectra you are given are generally simple ones, where isomerism is not a problem, so you should be able to identify each sample from its spectra.

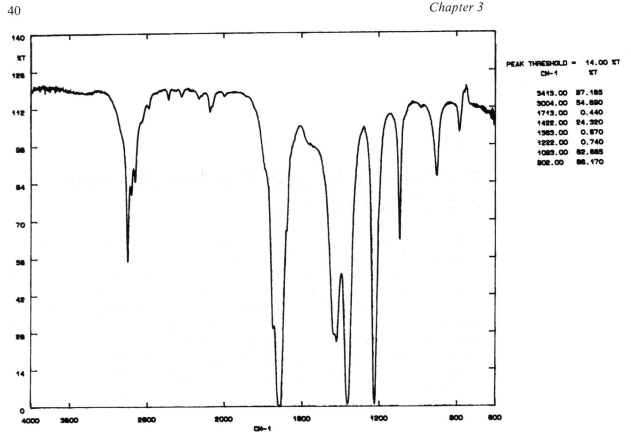

PEAK THRESHOLD = 14.00 %T
      CM-1   %T

     3413.00   97.185
     3004.00   54.890
     1713.00   0.440
     1422.00   24.320
     1383.00   0.870
     1222.00   0.740
     1083.00   82.885
      902.00   86.170

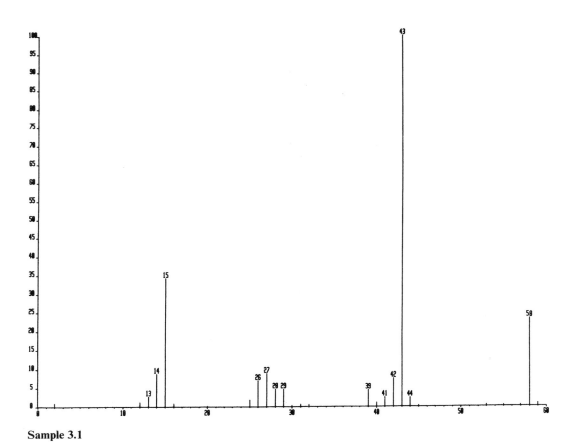

Sample 3.1

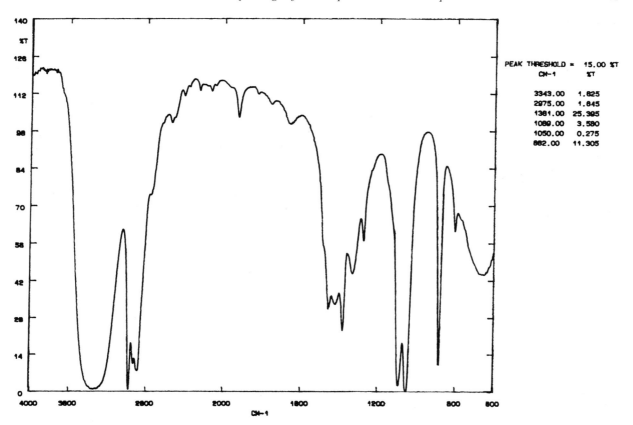

PEAK THRESHOLD = 15.00 %T

CM-1	%T
3343.00	1.825
2975.00	1.845
1381.00	25.395
1089.00	3.580
1050.00	0.275
882.00	11.305

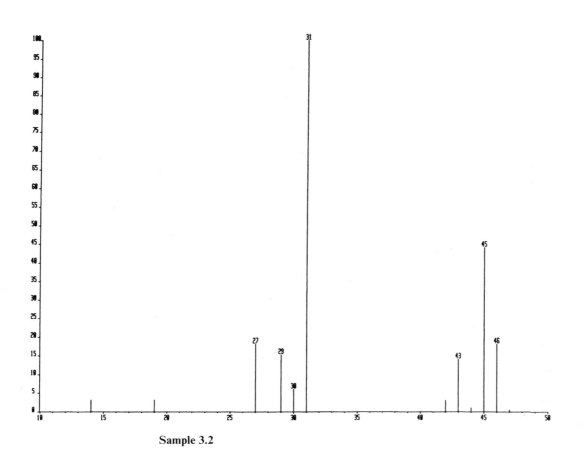

Sample 3.2

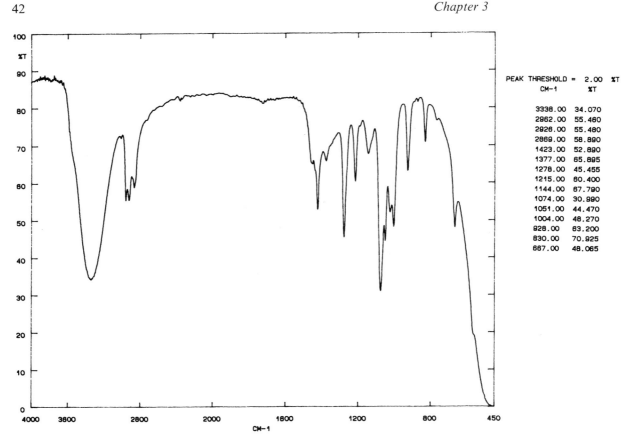

PEAK THRESHOLD =   2.00 %T
       CM-1   %T

    3338.00   34.070
    2962.00   55.460
    2926.00   55.480
    2869.00   58.890
    1423.00   52.890
    1377.00   65.895
    1278.00   45.455
    1215.00   60.400
    1144.00   67.790
    1074.00   30.990
    1051.00   44.470
    1004.00   48.270
     928.00   63.200
     830.00   70.925
     667.00   48.065

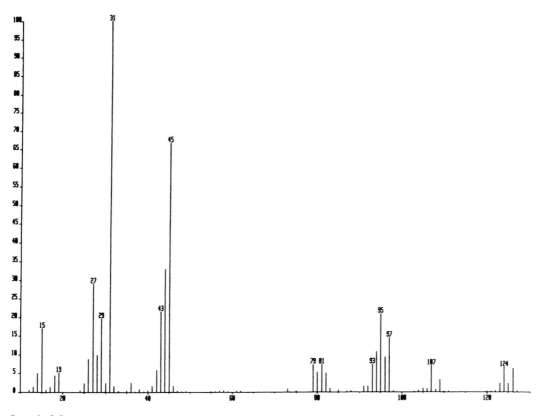

Sample 3.3

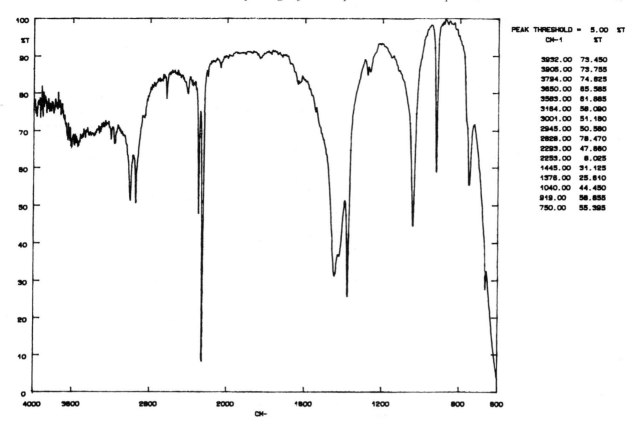

PEAK THRESHOLD = 5.00 %T

CM-1	%T
3932.00	73.450
3905.00	73.755
3794.00	74.825
3650.00	85.585
3583.00	81.885
3164.00	58.090
3001.00	51.180
2945.00	50.580
2826.00	78.470
2293.00	47.880
2253.00	8.025
1445.00	31.125
1376.00	25.610
1040.00	44.450
919.00	58.855
750.00	55.395

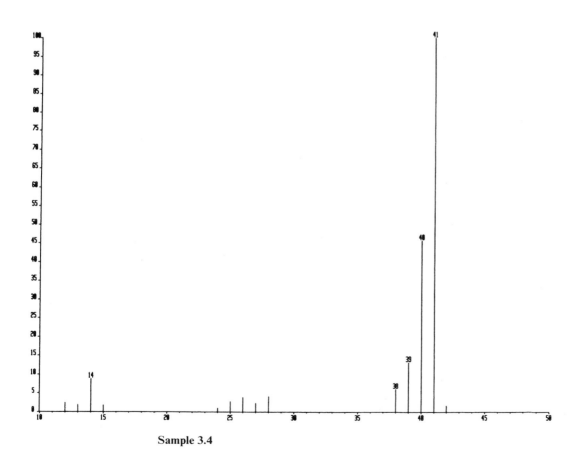

Sample 3.4

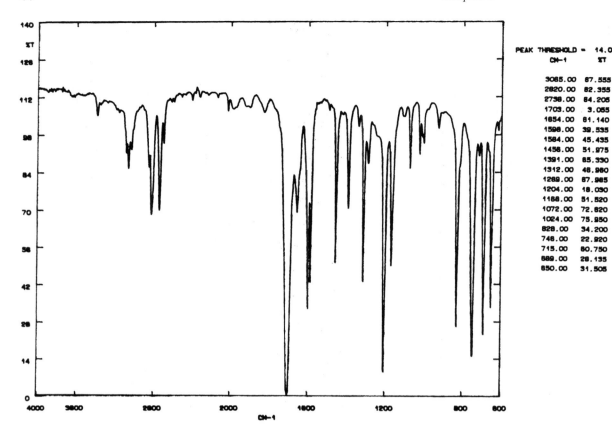

PEAK THRESHOLD =   14.00 %T
        CM-1   %T

        3085.00   87.555
        2820.00   82.355
        2738.00   84.205
        1703.00   3.055
        1654.00   81.140
        1598.00   39.535
        1584.00   45.435
        1456.00   51.975
        1391.00   85.330
        1312.00   48.980
        1269.00   87.985
        1204.00   18.090
        1188.00   51.520
        1072.00   72.820
        1024.00   75.950
         828.00   34.200
         748.00   22.920
         715.00   80.750
         689.00   28.135
         650.00   31.505

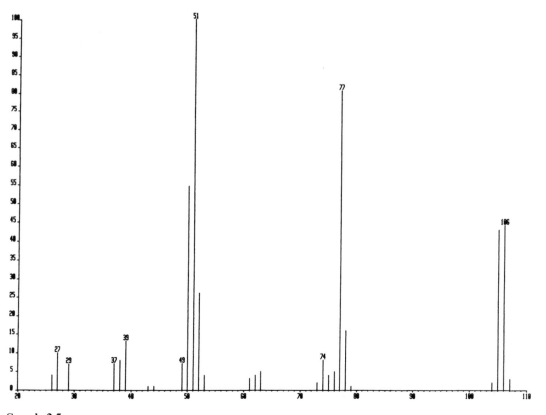

Sample 3.5

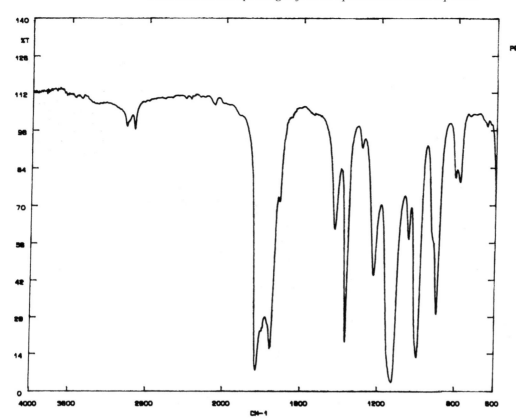

PEAK THRESHOLD = 2.80 %T

CM-1	%T
3569.00	111.900
2943.00	108.630
1827.00	12.420
1757.00	13.080
1430.00	35.070
1371.00	13.535
1294.00	75.520
1226.00	22.230
1126.00	11.245
1047.00	31.330
996.00	11.725
898.00	14.950
809.00	56.240
785.00	54.245
649.00	84.285

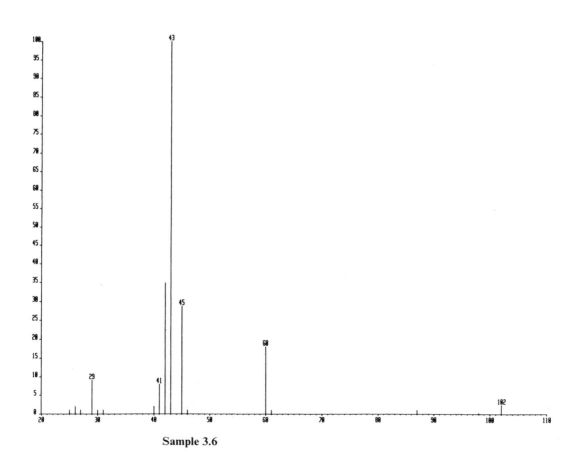

Sample 3.6

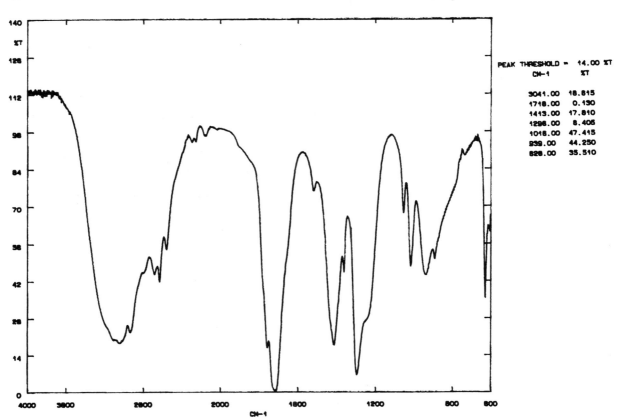

PEAK THRESHOLD = 14.00 %T

CM-1	%T
3041.00	18.815
1718.00	0.130
1413.00	17.810
1298.00	8.405
1018.00	47.415
939.00	44.250
828.00	35.510

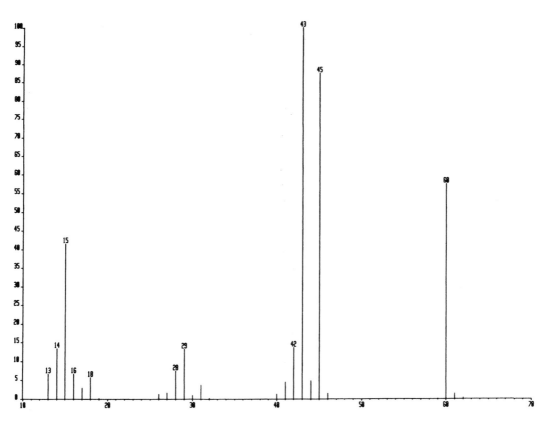

Sample 3.7

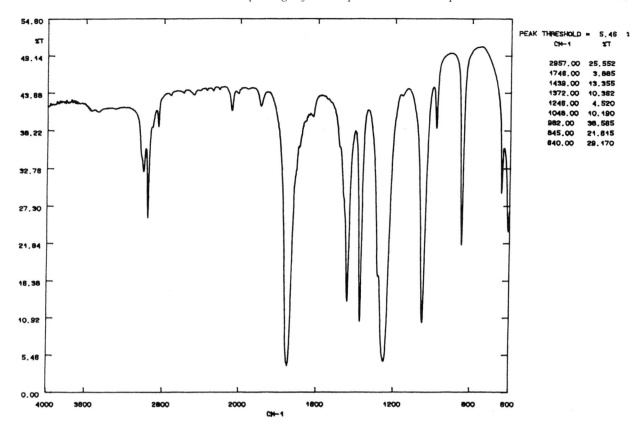

PEAK THRESHOLD = 5.46 %

CM-1	%T
2957.00	25.552
1746.00	3.885
1439.00	13.355
1372.00	10.382
1246.00	4.520
1048.00	10.190
982.00	38.585
845.00	21.815
640.00	29.170

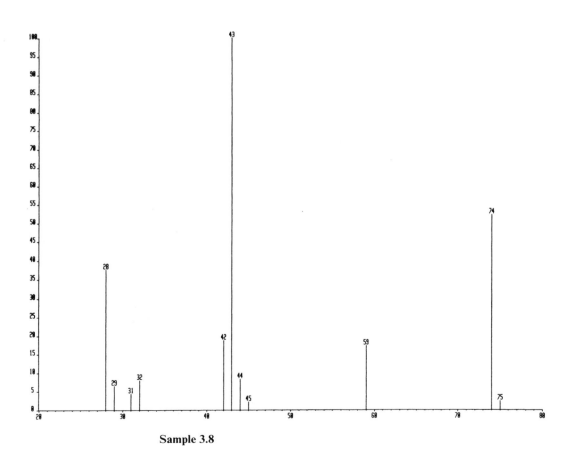

Sample 3.8

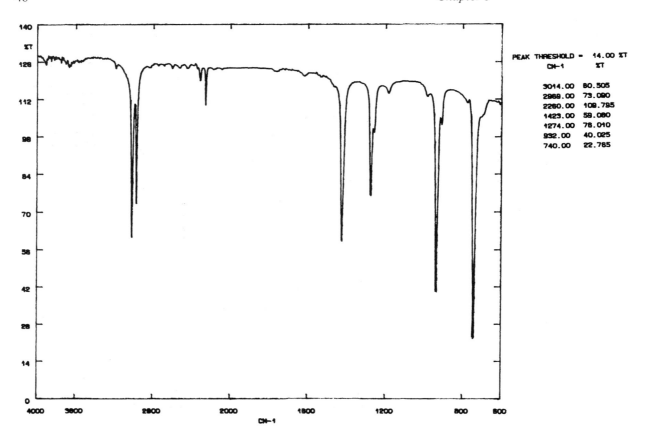

PEAK THRESHOLD =   14.00 %T
          CM-1   %T

       3014.00   80.505
       2968.00   73.090
       2260.00   109.795
       1423.00   59.080
       1274.00   78.010
        932.00   40.025
        740.00   22.785

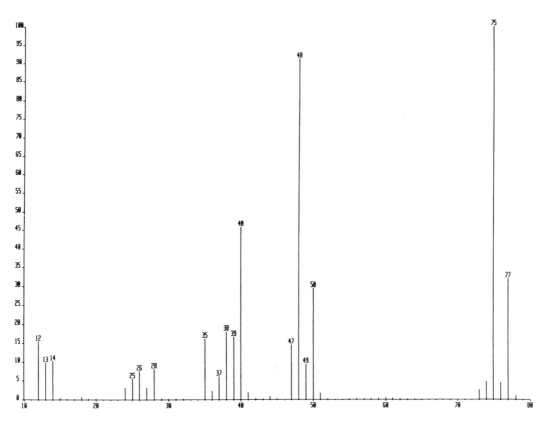

Sample 3.9

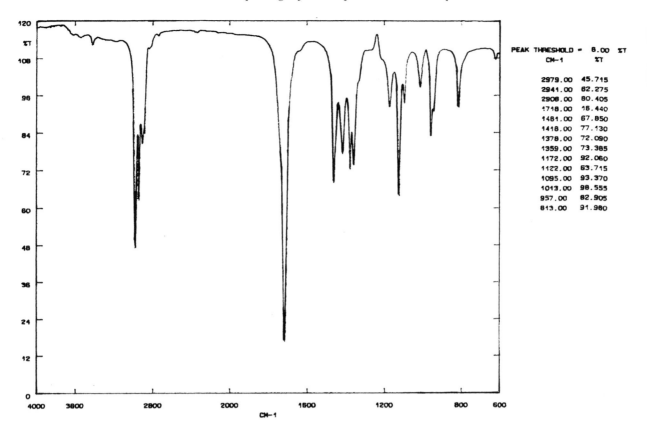

PEAK THRESHOLD =	6.00 %T
CM-1	%T
2979.00	45.715
2941.00	62.275
2908.00	80.405
1718.00	16.440
1461.00	67.850
1418.00	77.130
1378.00	72.090
1359.00	73.385
1172.00	92.060
1122.00	63.715
1095.00	93.370
1013.00	98.555
957.00	82.905
813.00	91.980

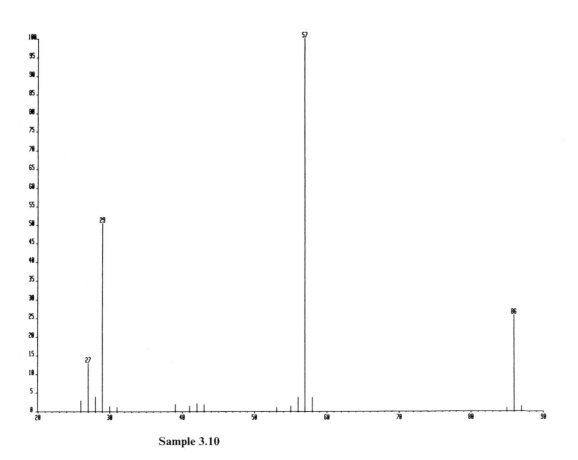

Sample 3.10

CHAPTER 4

Ultraviolet Spectroscopy

Ultraviolet spectroscopy tells us about conjugation in a molecule. It uses the highest energy radiation of any of the techniques described in this book; radiation at 200 nanometres is equivalent to 595 kilojoules per mole. Absorption of this energy raises an electron from a bonding orbital to an antibonding orbital.

Almost all ultraviolet spectroscopy involves radiation from 200 nm to 400 nm (1 nm = 10^{-7} cm = 10 Å = 1 mμ). The region is often extended from 400 to 800 nm to cover the visible spectrum. The region below 200 nm, often referred to as the far ultraviolet, requires special equipment, since oxygen absorbs just below 200 nm.

The ultraviolet spectrum measures the energy absorbed by an electron making the transition from a bonding to an antibonding orbital. If we look at a very simple, molecule, methane, CH_4, then we see that it contains only C–H bonds. Since it is symmetrical, all the C–H bonds are equivalent, and we need to consider only one bond. If we take a carbon atom and a hydrogen atom, and consider a single electron attached to each, as shown in Diagram 4.1, we have the energy levels shown. If we then bring these atoms together, and use the electrons to form a bond, then two orbitals are formed, these having differing energy levels.

Diagram 4.1 *Orbital changes when a C atom and an H atom form a C–H bond*

The bonding electrons occupy the lower level, known as the bonding orbital, denoted by σ. The upper orbital, known as the antibonding orbital and denoted by σ^*, is empty. When methane absorbs ultraviolet radiation, one of the bonding electrons can be raised to the antibonding orbital. The radiation at 140 nm has the right energy to make this transition occur. Consequently, methane does not absorb ultraviolet light between 200 and 400 nm. We can construct similar diagrams for C–C bonds, which have a $\sigma \rightarrow \sigma^*$ transition at 150 nm, and C=C bonds, where the $\pi \rightarrow \pi^*$ transition is at 170 nm.

When we consider a molecule which contains oxygen, we have to consider the effect of lone pairs of electrons, as well as σ and π bonding electrons. Thus, for the C=O bond of formaldehyde, $H_2C=O$, we can construct the diagram shown in Diagram 4.2. The $n \rightarrow \pi^*$ transition results from the absorption of energy at 270 nm, so this transition is found in the ultraviolet spectroscopy range. It is the only transition not involving d orbitals found in the ultraviolet range, as the $n \rightarrow \sigma^*$ (found in, for example, alcohols) absorbs radiation just above 200 nm.

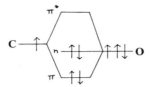

Diagram 4.2 *Orbital changes when a C atom and an O atom form a C=O bond*

On this basis, ultraviolet spectroscopy would appear to be of limited value to organic chemists, since n→π* transitions are quite rare. However, we have to consider conjugation. If we have a molecule with two ultraviolet absorbing groups, which we call chromophores, the observed spectrum will be the simple sum of the absorptions due to the two chromophores, unless we have conjugation. If we do, then the energy of the highest bonding orbital is raised, and that of the lowest antibonding orbital is lowered, as shown in Diagram 4.3. Consequently, the energy difference between the levels is lowered and the transition takes place at lower energy, and absorbs radiation of longer wavelength. The more chromophores are involved in the conjugation, the greater the effect.

Molecule	Absorption
Ethylene	170 nm
Butadiene	217 nm
Hexatriene	253 nm

Double bonds can also absorb radiation above 200 nm as a result of conjugation with alkyne, ketone, aldehyde, ester, nitrile or amide groups.

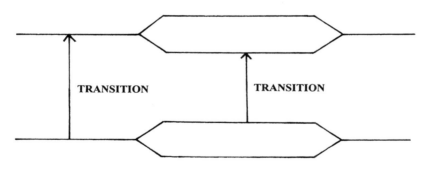

UNCONJUGATED   **CONJUGATED**

Diagram 4.3 *Changes in transition energy resulting from changes in the energy of bonding and antibonding orbitals as a result of conjugation*

The benzene ring, as might be expected, also absorbs ultraviolet radiation. Benzene itself absorbs at 184, 203.5 and 254 nm in hexane solution. The effect of conjugation with substituents is to move the absorption bands to longer wavelength. The band at 203.5 nm moves rapidly with increasing conjugation, and can overtake the band at 254 nm, which moves more slowly. The absorption can even move into the visible when we have an electron withdrawing and an electron attracting substituent in a 1,4- or *para*-substitution pattern, which is why compounds such as 1,4-nitrophenol are yellow.

A molecule which has absorbed ultraviolet radiation and been raised to an excited state can undergo many reactions which it would not undergo in the ground state. This field of chemistry is known as photochemistry. The excited state has a short lifetime, and usually undergoes decay by emitting its excess energy. However, some energy is lost by collisions during the excited state, so the energy emitted is always less than that absorbed; radiation is consequently emitted at a longer wavelength. If we choose a molecule which absorbs radiation of a slightly shorter wavelength than visible light, then it will emit energy in the visible region when irradiated with ultraviolet radiation, and appear to be a pale blue colour. This phenomenon is known as fluorescence. It can be seen if a glass of tonic water (which contains quinine) is taken out of doors into the sunshine, when it appears to be a very pale blue. Compounds of this type are added to most washing powders: a white item washed in one of these will retain some of the fluorescing agent; it then reflects most of the light falling on it, but absorbs some ultraviolet radiation and emits it in the visible range, so that more visible radiation comes from the item than falls on it. It can then be described as 'whiter than white'. This can be demonstrated by shining an ultraviolet light on to a shirt washed in any washing powder. In a dark room, the shirt glows with a pale blue light. The experiment requires great care, as ultraviolet light can rapidly damage eyes, and can burn skin exposed to it.

Ultraviolet spectra are measured by dissolving the sample in a suitable solvent, often hexane or methanol, and scanning it, measuring absorption as a function of wavelength. Two pieces of information should be obtained about any absorption peak. The first is the wavelength of the peak maximum, denoted by λ_{max}. The second is the intensity of the absorption. This depends on two laws, Lambert's law, which states that the fraction of the radiation absorbed is independent of the intensity of the radiation source, and Beer's law, which states that the absorption is proportional to the number of absorbing molecules. These laws lead to the equation

$$\log_{10} I_0/I = lc\varepsilon$$

where I_0 and I are the intensities of the incident and transmitted light, respectively, l is the path length of the cell in centimetres, and c is the concentration of the solution of the sample in moles per litre. The remaining factor, ε, is called the molar extinction coefficient for the particular peak of the compound being studied. By convention, it is always given without units (the actual units are $1000 \, cm^2 \, mol^{-1}$). The value of I_0/I is called the absorbance or optical density, and most spectrometers draw the spectrum as a graph of $\log_{10} I_0/I$ as a function of wavelength.

The value of ε_{max} (the molar extinction coefficient at the peak maximum) is useful because it tells us if the transition we are interested in is 'allowed' or 'forbidden'. Transitions are 'forbidden' for complex reasons of symmetry, but actually occur weakly because the symmetry which makes them forbidden is broken by molecular vibrations or by substitution. As an approximate rule, forbidden transitions have ε_{max} values below about 1000, usually well below. The n$\rightarrow\pi^*$ transition in ketones is an example of a forbidden transition, in which ε_{max} values are usually below 100. In contrast, $\pi\rightarrow\pi^*$ transitions are allowed, and ε_{max} is usually above 1000. Conjugation usually increases the value of ε_{max}.

Let us now look at an ultraviolet spectrum, shown in Figure 4.1. The spectrum is that of a saturated ketone, butan-2-one, $CH_3COCH_2CH_3$. The spectrum shows a single absorption peak at λ_{max} 279 nm, ε_{max} 16.6. This very low value of ε_{max} shows us that this is a forbidden transition, an n$\rightarrow\pi^*$ transition, characteristic of an aldehyde or ketone group or a nitro group. These peaks all occur within the general range of 275–290 nm. Note that

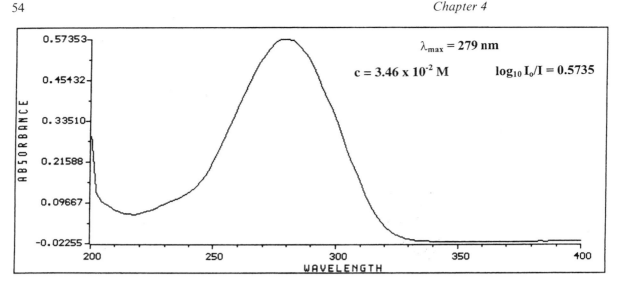

Figure 4.1

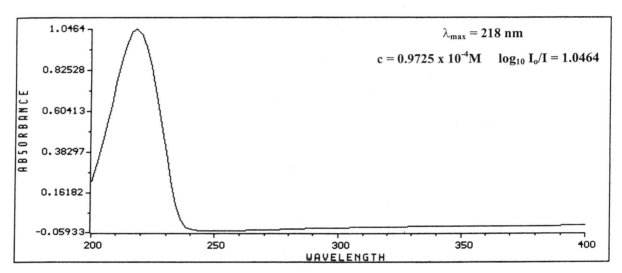

Figure 4.2

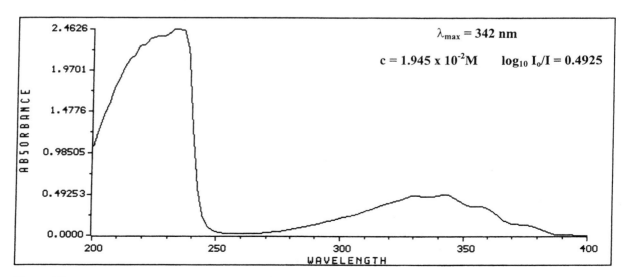

Figure 4.3

esters, acids and amides do not show this absorption; they absorb in the 200–215 nm region.

The next spectrum is that of an α,β-unsaturated ketone, cyclohexanone. The spectrum has two absorption peaks, of such widely differing ε_{max} values that they have to be recorded at different concentrations. The spectra obtained are shown in Figures 4.2 and 4.3. The more intense absorption is shown in Figure 4.2 which shows a peak at $\lambda_{max} = 218$, with $\varepsilon_{max} = 10\,760$. This is clearly not a forbidden transition, so must be the $\pi \rightarrow \pi^*$ transition moved to longer wavelength by conjugation. The weaker absorption is shown in Figure 4.3 The region around 200–250 nm is swamped by the intense absorption of the $\pi \rightarrow \pi^*$ transition, but the spectrum shows a peak at $\lambda_{max} = 342$ nm, $\varepsilon_{max} = 25.3$. The low intensity of this peak shows it to be a forbidden transition, an $n \rightarrow \pi^*$ transition. It is at higher wavelength than the previous $n \rightarrow \pi^*$ transition and has slightly higher ε_{max} as a result of conjugation with the $\pi \rightarrow \pi^*$ transition. Conjugated $n \rightarrow \pi^*$ carbonyl peaks usually occur in the 300–350 nm region, and have ε_{max} below 100. Clearly, we have a double bond conjugated with a carbonyl group.

Summary

The first thing to do with an ultraviolet spectrum is to observe the values of λ_{max} and $\log_{10} I_0/I$ at λ_{max}, and hence calculate ε_{max}.

Then, if ε_{max} is over about 1000, we have an allowed transition, usually $\pi \rightarrow \pi^*$, and the value of λ_{max} gives some idea as to the amount of conjugation involved. The exact position of the peak can provide a lot of information about the conjugated system, using a complex set of rules known as Woodward's rules.

If ε_{max} is below about 1000, then we have a forbidden transition, usually $n \rightarrow \pi^*$. Peaks in the range 275–295 nm are not conjugated; peaks in the range 300–350 nm are conjugated to a $\pi \rightarrow \pi^*$ transition. Conjugation of two carbonyl peaks can occur in dicarbonyl compounds, and leads to two $n \rightarrow \pi^*$ transitions at 290 nm and 340–440 nm, which is why some of these compounds are yellow.

Problems in Interpreting Ultraviolet Spectra

You are given the ultraviolet spectra of five samples. Calculate the ε_{max} value of each peak, and hence decide if the peak is $\pi \rightarrow \pi^*$ or $n \rightarrow \pi^*$. Use the λ_{max} values to decide if either the $\pi \rightarrow \pi^*$ or the $n \rightarrow \pi^*$ transition involves a conjugated system. Hence, suggest the type of structure, *e.g.* diene or ketone or α,β-unsaturated ketone, present in the sample.

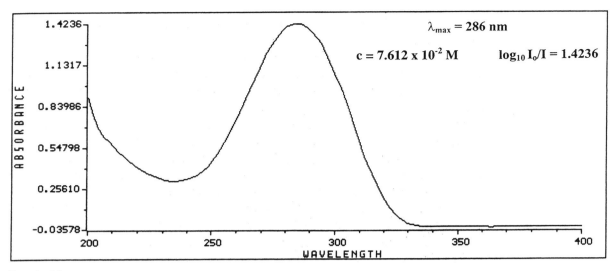

$\lambda_{max} = 286$ nm

$c = 7.612 \times 10^{-2}$ M $\log_{10} I_o/I = 1.4236$

Sample 4.1

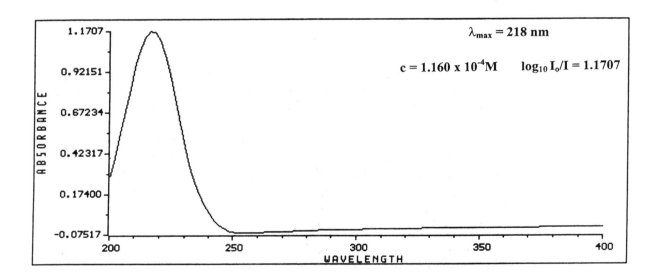

$\lambda_{max} = 218$ nm

$c = 1.160 \times 10^{-4}$M $\log_{10} I_o/I = 1.1707$

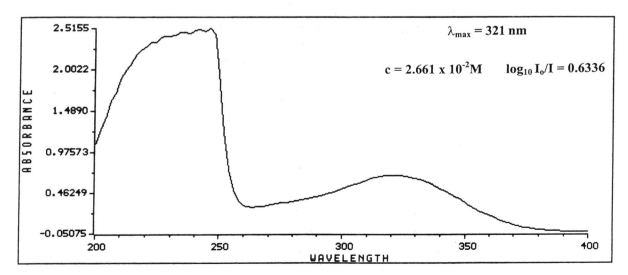

$\lambda_{max} = 321$ nm

$c = 2.661 \times 10^{-2}$M $\log_{10} I_o/I = 0.6336$

Sample 4.2

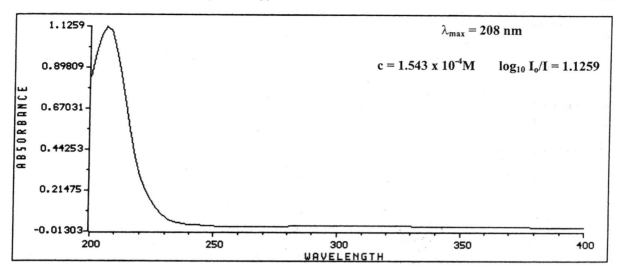

$\lambda_{max} = 208$ nm

$c = 1.543 \times 10^{-4}$M $\log_{10} I_o/I = 1.1259$

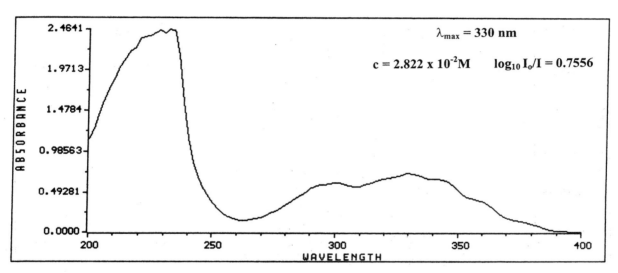

$\lambda_{max} = 330$ nm

$c = 2.822 \times 10^{-2}$M $\log_{10} I_o/I = 0.7556$

Sample 4.3

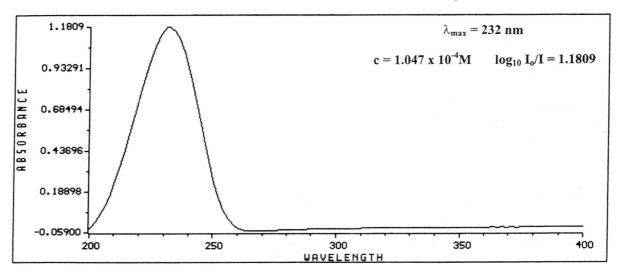

$\lambda_{max} = 232$ nm

$c = 1.047 \times 10^{-4} M$ $\log_{10} I_o/I = 1.1809$

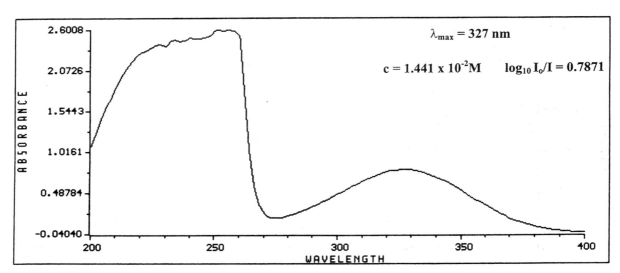

$\lambda_{max} = 327$ nm

$c = 1.441 \times 10^{-2} M$ $\log_{10} I_o/I = 0.7871$

Sample 4.4

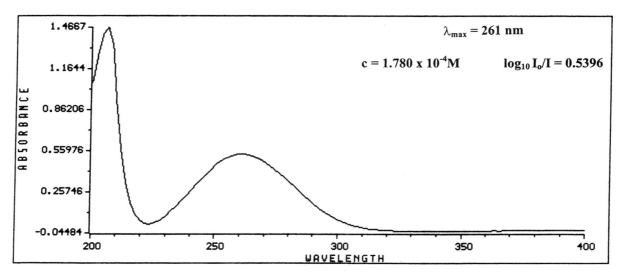

$\lambda_{max} = 261$ nm

$c = 1.780 \times 10^{-4} M$ $\log_{10} I_o/I = 0.5396$

Sample 4.5

CHAPTER 5

^{13}C Nuclear Magnetic Resonance Spectroscopy

^{13}C NMR spectroscopy tells us about the environments of the carbon atoms in a molecule. The technique of NMR can be applied only to atoms which have nuclear spin; when they are placed in a strong magnetic field, the nuclear spins can take up different orientations which are at different energy levels. For nuclei with nuclear spin of ½, such as ^{1}H, ^{13}C, ^{19}F and ^{31}P, the nuclei orient themselves in one of two ways*. When the nucleus is irradiated with radiation of the appropriate frequency (radiofrequency), some of the nuclei in the lower energy state absorb radiation and are raised to the higher energy state. When irradiation stops, they relax to the equilibrium distribution. This relaxation process is quite slow.

The frequency of the radiation absorbed depends on the magnetic field applied to the nucleus. The field is normally made up of three components:

A. The externally applied field. This is an extremely powerful magnetic field. When a field of 23.49 kilogauss is applied to a carbon nucleus, it absorbs radiation at 25.14 MHz. The same field, applied to a hydrogen nucleus, would make it absorb radiation at 100 MHz. Since most machines run ^{13}C and ^{1}H NMR spectra, the radiation absorbed by the latter is often used to describe the power of the machine. Thus, a machine with a field of 23.49 kilogauss would be described as a 100 MHz spectrometer. To attain high, stable magnetic fields, superconducting magnets, kept at the temperature of liquid helium, are often used. The externally applied field is responsible for almost all the magnetic field applied to the nucleus, but the field is slightly modified by the effects of neighbouring atoms. If this did not happen, all carbon atoms would absorb radiation at the same frequency.

B. The applied field can be affected by the variation in electron density around the neighbouring nuclei. Movement of these electrons in the applied field creates a magnetic field which opposes the applied field, hence changing the absorption frequency. Since hydrogen is (almost) the least electronegative element, replacement of a hydrogen on a carbon atom by any other atom moves the absorption frequency of that carbon atom to that associated with a lower magnetic field, usually described as downfield. This movement is referred to as chemical shift, and it is always proportional to the applied field.

C. The applied field is affected by the field resulting from the nuclear spin of neighbouring nuclei. These can be aligned with or against the applied field, either increasing or decreasing it. Consequently, a nucleus having one neighbouring atom with nuclear spin has an equal chance of being

---

*They are often pictured as being like small bar magnets, and aligning themselves with or against the magnetic field—a useful, though simplistic, illustration.

in a field increased or decreased by that neighbouring atom, and hence the absorption peak is split into two equal peaks. This interaction of nuclei, described as coupling, results directly from the magnetic field of the nucleus, so is not influenced by the strength of the applied field.

For any nucleus, the frequency of the radiation absorbed depends on the magnetic field, so that we have two variables to consider when constructing a scale. It is difficult to measure very high magnetic fields with the degree of accuracy required to produce an absolute scale, so absorption frequencies are measured as the frequency in Hertz from an arbitrary standard. In ^{13}C and ^{1}H NMR spectroscopy this standard is tetramethylsilane [TMS, $Si(CH_3)_4$], which absorbs upfield of most carbon and hydrogen atoms. The position of absorption of any peak depends on the applied field, so that an absorption peak which is at 50 Hz from TMS in a 100 MHz spectrometer will be at 100 Hz from TMS in a 200 MHz spectrometer. We therefore measure absorption frequencies on the δ scale, defining δ as:

$$\delta = \frac{\text{Shift in Hz from TMS}}{\text{Operating frequency in MHz}}$$

Thus, in our example

$$\delta = \frac{50}{100} = \frac{100}{200} = 0.5$$

The value of δ has no units, and is expressed as fractions of the applied field in parts per million (ppm).

To measure a ^{13}C NMR spectrum, we place a solution of the compound being studied in a magnetic field, spin it to even out the effect of any variations in the field, then irradiate it and measure the absorption of radiation as a function of the frequency of radiation.

The permanent problem of ^{13}C NMR spectroscopy is that natural carbon is only 1.1% ^{13}C. Thus, even a pure liquid is only a dilute solution of the material which is giving a spectrum, and background electronic noise levels are high. One possible approach would be to scan the spectrum repeatedly and accumulate the spectra on top of each other, so that electronic noise cancelled out, and the spectrum emerged. This process would be extremely slow, since each scan takes about five minutes. The problem has been solved by using rapidly repeated scans with a single powerful pulse covering the whole frequency range. This can be repeated until a sufficiently strong spectrum is obtained, which is then transformed into the spectrum obtained from a conventional scan by a Fourier transform. The method is highly efficient, but has one major snag. Since the sample is scanned many times in rapid succession, not all nuclei have time to revert to their equilibrium distribution of nuclear spin states before the next burst of scanning radiation comes along. Consequently the nuclei with longer relaxation times, which are not fully relaxed, absorb less energy from the second and subsequent bursts of radiation than do those nuclei which are fully relaxed. As a result, the absorption band in the final spectrum from an atom with a long relaxation time, such as a carbonyl or quaternary carbon atom, is smaller than the absorption band from an atom with a shorter relaxation time. As a result, in a Fourier Transform (FT) spectrum, we cannot use peak areas as a measure of the number of nuclei in a particular environment.

We can, however, make good use of this phenomenon. Most NMR spectra are measured in solution. We need a solvent which, ideally, does not contain

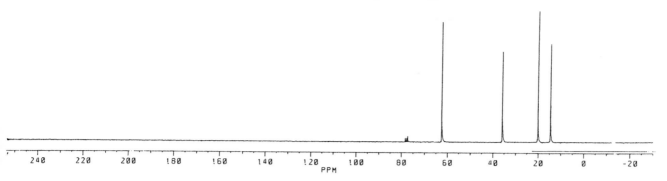

1-BUTANOL NON-DECOUPLED

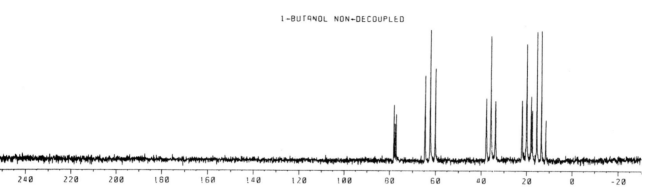

Figure 5.1

any carbon atoms. This is not practical in ^{13}C NMR work, but if we choose a solvent in which the carbon atoms have a deuterium substituent, then that solvent will have a very weak ^{13}C absorption spectrum because a deuterium substituent greatly lengthens the relaxation time of a carbon atom. We choose CDCl$_3$ for most purposes. Since the deuterium nucleus has a spin of 1, it splits the carbon peak into three equal peaks, so our solvent gives three relatively weak peaks around $\delta 77$.

An example of a ^{13}C NMR spectrum is shown in Figure 5.1. It shows the ^{13}C NMR spectrum of butan-1-ol, CH$_3$–CH$_2$–CH$_2$–CH$_2$–OH. The figure shows two traces. The upper is known as the decoupled spectrum, and shows one line for each carbon atom in a different environment. In this case, there are four lines, since all the four carbon atoms are in different environments, and hence have different chemical shifts. The lower trace shows the undecoupled spectrum, in which each of these peaks is split into a multiplet by the attached hydrogen atoms.

We have already seen that the absorption band from an atom is split into two bands by the magnetic field of a neighbouring nucleus with nuclear spin. This is not a problem with ^{13}C to ^{13}C interaction, since ^{13}C is present to only 1.1%, so any ^{13}C atom is likely to have only ^{12}C neighbours, which do not have nuclear spin. The small amount of splitting from ^{13}C to ^{13}C coupling would be lost in the electronic noise. However, the signal from any ^{13}C atom with a ^{1}H atom would be split by the magnetic field of that atom. In order to obtain the upper trace, we observe the ^{13}C absorption while strongly irradiating the sample with radiation absorbed by ^{1}H atoms. This causes the ^{1}H atoms to flip very rapidly between spin states several times while the ^{13}C signal is being recorded. The ^{13}C atom thus experiences an averaged spin system from the neighbouring ^{1}H nucleus, and is not split. This process is referred to as decoupling, and the ^{13}C spectrum recorded under these conditions is called a decoupled spectrum.

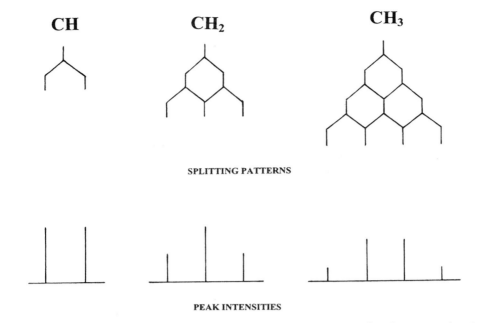

SPLITTING PATTERNS

PEAK INTENSITIES

Diagram 5.1 *Splitting patterns and peak intensities for the spectra of carbon atoms bearing hydrogen atoms*

If we then turn the irradiation off, then the carbon signals are split by the effects of the hydrogen atoms, and we get the lower trace shown in Figure 5.1. This shows the ^{13}C to ^{1}H splitting, and enables us to say how many hydrogen atoms are attached to each carbon atom.

If the carbon atom does not have any attached hydrogen atoms, it remains a singlet. If it has one attached hydrogen atom, it is split into a doublet. If it has two attached hydrogen atoms, then two equal splittings give a triplet with peaks in the ratio $1:2:1$. Since having n attached hydrogens splits the peak into $(n + 1)$ peaks, this is known as the $(n + 1)$ rule (Diagram 5.1). We can now tell how many different carbon atoms the molecule has, and how many hydrogen atoms are attached to each carbon atom.

This method works well in a molecule which has well spaced ^{13}C peaks, but is less successful in larger molecules. An alternative method is to measure what it known as a DEPT spectrum (Distortionless Enhancement by Polarisation Transfer). The DEPT spectrum plots the CH_3, CH_2 and CH peaks separately. It does not record carbon atoms which do not have a hydrogen atom attached, so we need a decoupled spectrum as well as a DEPT spectrum. The spectra for butan-1-ol are shown in Figure 5.2.

The ^{13}C NMR spectrum of butan-1-ol thus consists of four peaks (+ $CDCl_3$ and TMS), $\underline{C}H_3$ at $\delta14$, and $\underline{C}H_2$ at $\delta19$, 35 and 63. Their δ values depend on the distance from the electronegative oxygen atom. Furthest downfield at $\delta63$ is the CH_2 group attached to the OH group. The influence of this oxygen is strong, and also affects the next CH_2 at $\delta35$, but the next CH_2 at $\delta19$ is almost unaffected. The CH_3 is at $\delta14$.

Almost all ^{13}C NMR peaks are found downfield from TMS, extending to about 200δ. Exact chemical shifts depend on the group itself, and on neighbouring electronegative groups. Thus, alkyl substitution of a CH_3 group, making it a CH_2 group, moves the absorption at $\delta8.4$ down to $\delta15.9$; a second substitution makes it a CH group at $\delta25.0$. The most useful chemical shifts are listed in Table 5.1. It is also useful to have the ranges of chemical shifts of the substituted alkyl groups, and these are given in Table 5.2.

DECOUPLED SPECTRUM

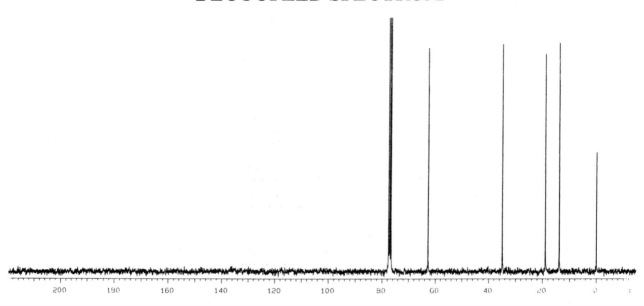

DEPT SPECTRUM

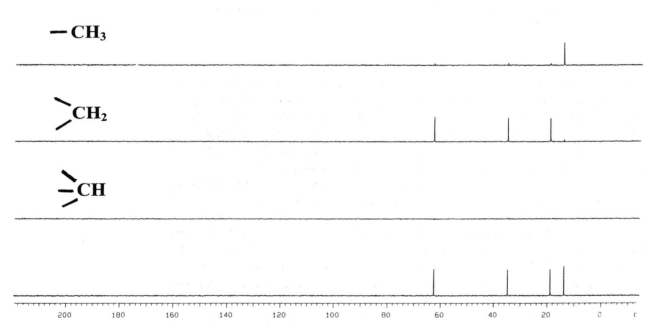

Figure 5.2

The next spectrum, shown in Figure 5.3, is that of butan-2-ol, CH_3–$CH(OH)$–CH_2–CH_3. Again, the decoupled spectrum consists of four peaks, so all carbon atoms are different. The DEPT spectrum shows that we have a C̲H at δ69, a C̲H_2 at δ32 and two CH_3 groups at δ23 and 10. The peak at δ69 is clearly that of the carbon atom attached to the oxygen, on the grounds of both chemical shift and splitting. The CH_3 group at δ23, being downfield of the other CH_3, must be the one attached to the CH(OH). The molecule is thus

$$CH_3\text{–}CH(OH)\text{–}CH_2\text{–}CH_3$$
$$\delta \quad 23 \quad 69 \qquad 32 \quad 10$$

Table 5.1 ^{13}C *NMR chemical shifts*

Group	^{13}C chemical shift δ (ppm)	
	Attached to groups of low electronegativity	Full range
CH_3	8	5–70
CH_2	16	10–80
CH	25	20–90
C≡C	67	65–80
C≡N	118	115–130
C=C	123	100–170
Ph	128	110–170
C=O ester	170	160–180
C=O acid	178	170–190
C=O ketone or aldehyde	200	190–210

Table 5.2 *Effect of electronegative groups on* ^{13}C *chemical shifts*

Group	Chemical shift δ (ppm) when next to				
	Alkyl	C=O	C=C	Ph	−OR*
CH_3	8–30	20–35	20–30	20–45	55–70
CH_2	15–40	25–40	20–40	35–55	50–80
CH	20–50	30–40	25–40	40–60	65–90

*R = H or alkyl.

The peaks from the carbon atoms of an aromatic ring show a strong downfield shift, as shown in Table 5.1. In the case of the ^{13}C NMR spectrum of methylbenzene (toluene), $PhCH_3$, shown in Figure 5.4, the aromatic carbon peaks are in the range of $\delta120$–140. Furthest downfield is the carbon attached to the methyl group. The other carbon atoms are moved downfield by amounts depending on their position relative to the methyl group. The methyl group itself is at $\delta21$, moved downfield by the phenyl ring.

We can now tackle a few spectra of unknown molecules. The first is shown in Figure 5.5. The substance has the molecular formula $C_4H_8O_2$, and has ^{13}C NMR peaks at $\delta171$, 60, 21 and 14. Thus, all four carbon atoms show separately on the decoupled spectrum. The peak at $\delta171$ does not show in the DEPT spectrum, so is a quaternary carbon atom. The peak at $\delta60$ is a CH_2 carbon atom, while the peaks at $\delta21$ and 14 are CH_3 groups. We thus have a CH_2 and two CH_3 groups, accounting for all eight hydrogen atoms. We also have only two 'chain ending' groups (groups which can bond to only one other group; in this case, the two CH_3 groups) so the molecule cannot have a branched chain.

We can account for C_4H_8 from the spectrum, out of $C_4H_8O_2$, so have to consider two oxygen atoms. The chemical shift of the quaternary carbon atom is $\delta171$, suggesting it is part of an ester linkage. The chemical shift of the CH_2 at $\delta60$ suggests it is attached to the ester oxygen. The more downfield of the CH_3 groups is attached to the carbonyl oxygen. The molecule is thus ethyl acetate:

$$CH_3\text{–}CO\text{–}O\text{–}CH_2\text{–}CH_3$$
$$\delta \quad 21 \quad 171 \quad\quad 60 \quad 14$$

The next unknown has the spectrum shown in Figure 5.6. The molecule is C_5H_8O, and it has ^{13}C NMR peaks at $\delta220$, 38 and 23. In this case, the

DECOUPLED SPECTRUM

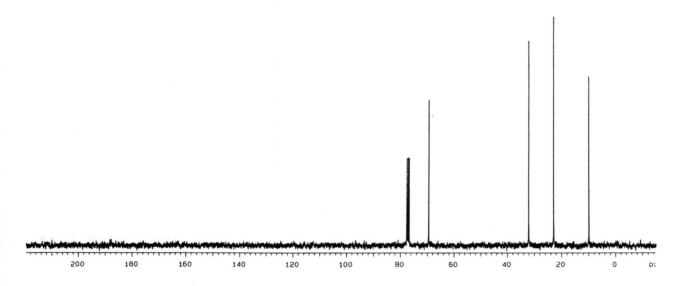

DEPT SPECTRUM

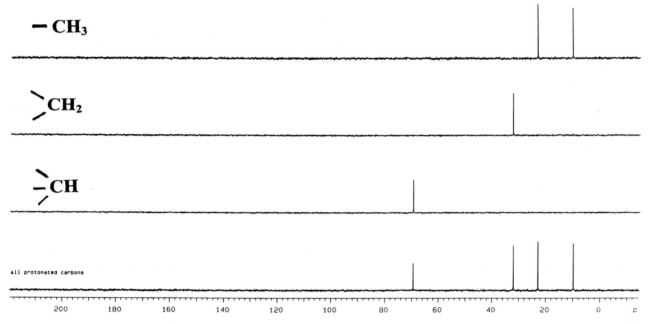

Figure 5.3

molecule has enough symmetry to put two pairs of atoms in identical environments. The peak at $\delta 220$ does not show in the DEPT spectrum, so is a quaternary carbon atom. The peaks at $\delta 38$ and 23 are both CH_2 groups, which accounts for only four out of eight hydrogen atoms. The molecule must thus consist of a quaternary carbon atom, two pairs of identical CH_2 groups, and one remaining oxygen. It is noticeable that the molecule does not have any chain ending groups, so it must be cyclic.

On the basis of its chemical shift, the quaternary carbon at $\delta 220$ is probably part of a ketone carbonyl group, which accounts for the oxygen atom. Chemical shifts also indicate that the carbon atoms with $\delta 38$ are closer to the

DECOUPLED SPECTRUM

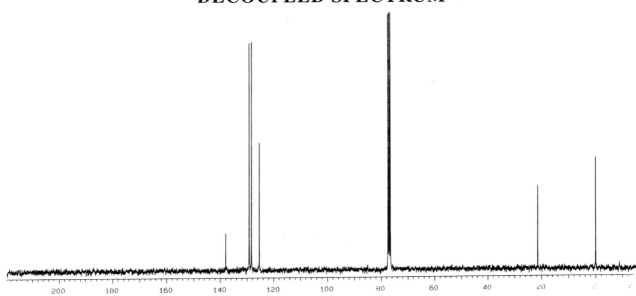

DEPT SPECTRUM

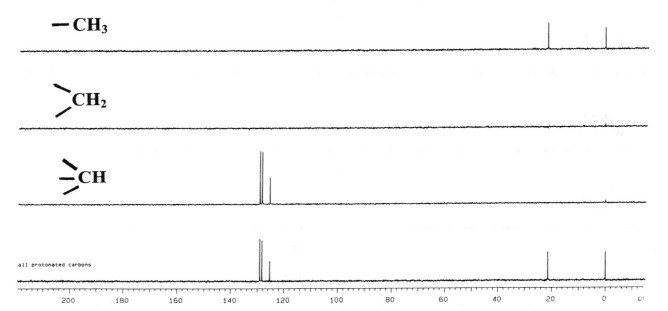

Figure 5.4

carbonyl group than those with $\delta 23$. The molecule is thus cyclopentanone:

$$
\begin{array}{ll}
& \text{O} \\
& \| \\
& \text{C} \\
\text{H}_2\text{C} \diagdown \diagup \text{CH}_2 \\
\text{H}_2\text{C} - \text{CH}_2
\end{array}
\quad
\begin{array}{l}
\delta\,220 \\
\delta\,38 \\[6pt]
\delta\,23
\end{array}
$$

The next unknown, whose spectrum is shown in Figure 5.7, has a molecular formula C_6H_{12}. The ^{13}C NMR spectrum has peaks at $\delta 139$ (CH), 114 (CH_2), 34 (CH_2), 31 (CH_2), 22 (CH_2) and 14 (CH_3). We can thus account for all six carbon atoms, all in different environments, and all 12 hydrogen atoms. On

DECOUPLED SPECTRUM

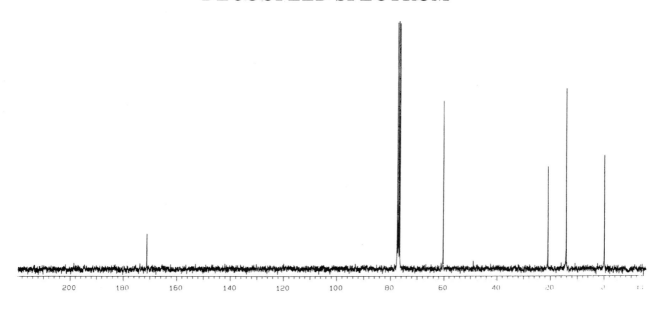

DEPT SPECTRUM

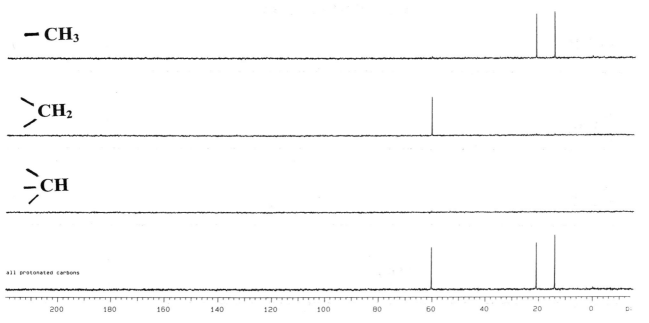

Figure 5.5

the basis of chemical shift, the carbon atoms at $\delta 139$ and 114 are probably atoms of a carbon double bond; since the latter is a CH_2 group, then the group is $-CH=CH_2$. This group and the methyl group provide the only chain ending groups, so the molecule must be a straight chain hydrocarbon, hex-1-ene:

$$CH_2=CH-CH_2-CH_2-CH_2-CH_3$$
$$\delta \quad 116 \quad 139 \quad 34 \quad 31 \quad 22 \quad 14$$

It is easy to assign a straight chain structure like this, but if we had an extra methyl group on C-3 or C-4 it would be difficult to be sure which carbon atom

DECOUPLED SPECTRUM

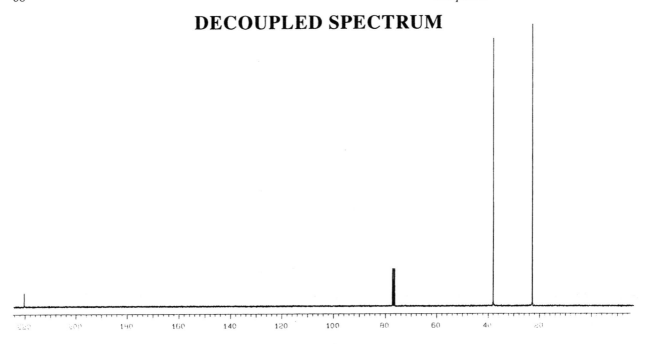

DEPT SPECTRUM

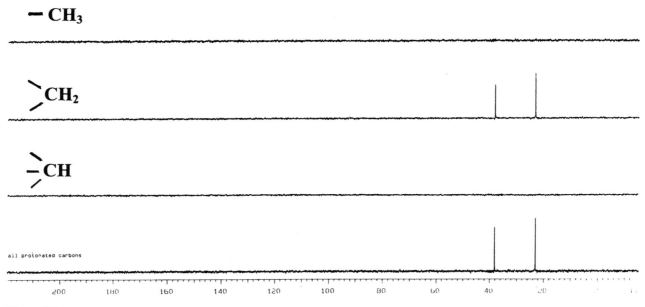

Figure 5.6

it was attached to. A methyl group on C-5 would be identical to C-6, and easy to assign. It is a general difficulty of ^{13}C assignment that branched chains can be difficult to assign unless symmetry considerations help.

The next unknown, shown in Figure 5.8, has the molecular formula C_8H_{16}. The spectrum has peaks at $\delta 130$ (CH), 35 (CH_2), 23 (CH_2) and 14 (CH_3). This accounts for only C_4H_8, suggesting that the molecule has a plane of symmetry, with two C_4H_8 units, one each side of the plane. The chemical shift of the peak at $\delta 130$ suggests that it is part of a double bond. Since it is only half of a double bond, the plane of symmetry must bisect the double

DECOUPLED SPECTRUM

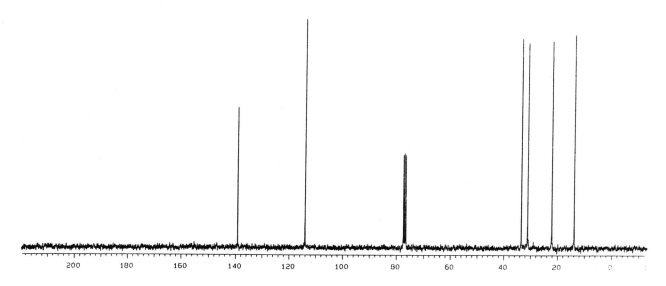

DEPT SPECTRUM

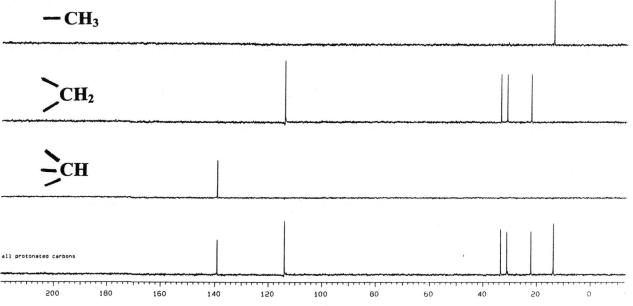

Figure 5.7

bond. The whole molecule has only two chain ending groups, so is linear. It is clearly oct-4-ene:

$$CH_3–CH_2–CH_2–CH=CH–CH_2–CH_2–CH_3$$
$$\delta \quad 14 \quad 23 \quad 35 \quad 130 \quad 130 \quad 35 \quad 23 \quad 14$$

We cannot, on the basis of this single spectrum, say whether the double bond is *cis* or *trans* (*E* or *Z*) (it is actually *trans*). In some cases, we could not distinguish the isomers even if we had spectra of authentic samples available. This is a restriction of ^{13}C NMR, though the problem can often be solved by ^1H NMR.

Our final example has the spectrum shown in Figure 5.9, and has the molecular formula $C_5H_{10}O$. It has peaks at $\delta 136$ (C), 124 (CH), 59 (CH_2), 26

DECOUPLED SPECTRUM

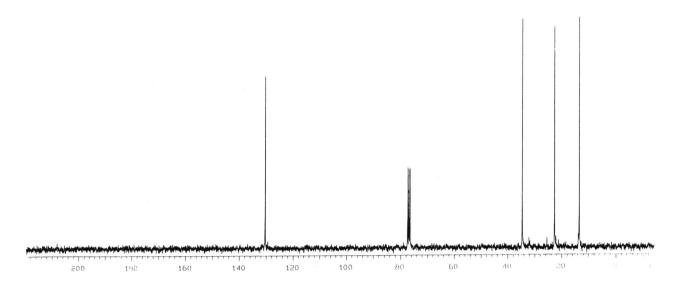

DEPT SPECTRUM

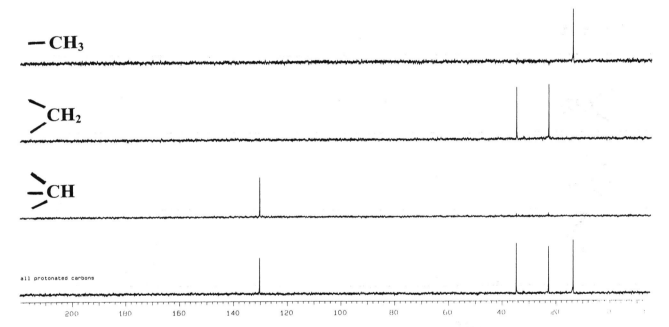

Figure 5.8

(CH₃) and 18 (CH₃). This accounts for five carbon atoms and nine hydrogen atoms, leaving OH, which suggest the molecule is an alcohol. Chemical shifts show the peaks at δ136 and 124 to be part of a double bond, HC=C, and the peak at δ59 is a CH₂ which must be attached to the OH group. The molecule has three chain ending groups, CH₃, CH₃ and CH₂OH, so it cannot be linear; these three groups must be attached to the double bond. We can do this in two ways:

DECOUPLED SPECTRUM

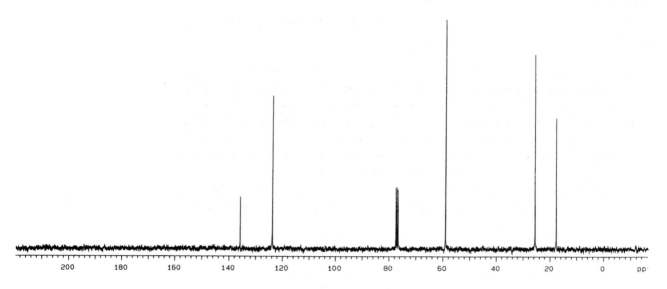

DEPT SPECTRUM

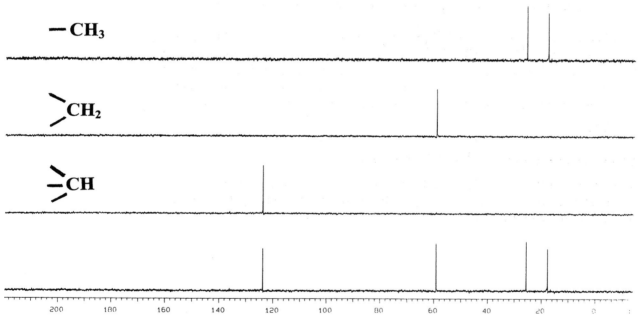

Figure 5.9

We cannot distinguish between these possible structures on the basis of the spectrum in Figure 5.9. Having spectra of I and II should allow us to make the distinction (Figure 5.9 is actually the spectrum of 3-methylbut-2-en-1-ol, I). This highlights the main problem of interpreting ¹³C NMR spectra: we cannot always tell which carbon atom is next to which. In simple molecules, it may not matter, or the effect of a strong electronegative group may help. Symmetry is also useful, but many large molecules have ¹³C spectra which cannot be easily assigned. The problem can be solved by ¹H NMR.

Summary

When we have a ^{13}C NMR spectrum to assign, the following steps are a useful approach:

A. Count the number of peaks in the decoupled spectrum. Compare the figure with the molecular formula. If you have more carbon atoms than peaks, then at least one peak represents more than one carbon atom.

B. Calculate the total number of hydrogen atoms on the carbon atoms, and see if this fits with the number of hydrogen atoms in the molecule. When you have allowed for identical groups, any hydrogen left must be attached to an element other than carbon.

C. See if any of the carbon atoms shown in the spectrum are part of recognisable groups, such as carbonyl groups, aromatic rings, double bonds, nitrile groups, *etc.*

D. See how many chain ending groups (*i.e.* groups with only one attachment to other atoms remaining) the molecule has. Zero implies a ring, one a substituted ring, two a straight chain or disubstituted ring and more than two a branched chain or polysubstituted ring.

E. Look at the molecular formula for atoms such as O, N, S, halogen. They can give useful clues as to the cause of large downfield shifts.

F. Relate chemical shifts to electronegative groups in the molecule.

In complex spectra, it is often possible to suggest several structures consistent with a ^{13}C NMR spectrum. Some of these may be eliminated on the grounds of symmetry, but in some cases a precise identification is not possible. The ^{1}H NMR spectrum will help in many cases, but cyclic systems have very complex ^{1}H NMR spectra and are more easily identified from ^{13}C spectra.

Problems in Interpreting ^{13}C NMR Spectra

You are given the decoupled and DEPT spectra of 10 samples, together with the molecular formula. In all cases, you should be able to identify the structure of the sample from the data provided.

DECOUPLED SPECTRUM
C_2H_6O

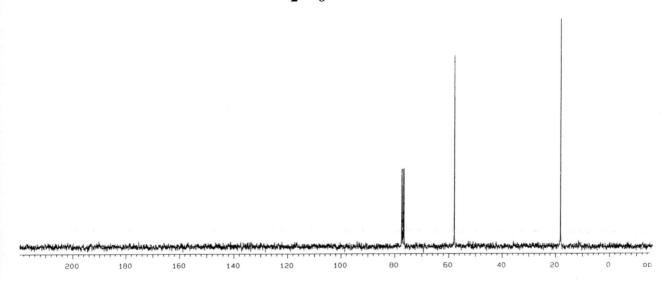

DEPT SPECTRUM

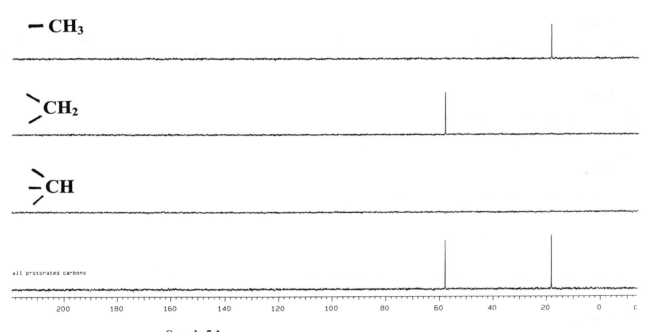

— CH₃

CH₂

CH

all protonated carbons

Sample 5.1

DECOUPLED SPECTRUM
$C_3H_6O_2$

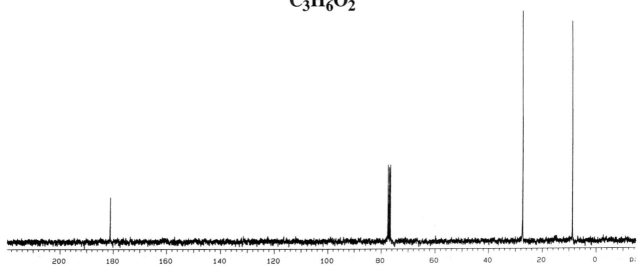

DEPT SPECTRUM

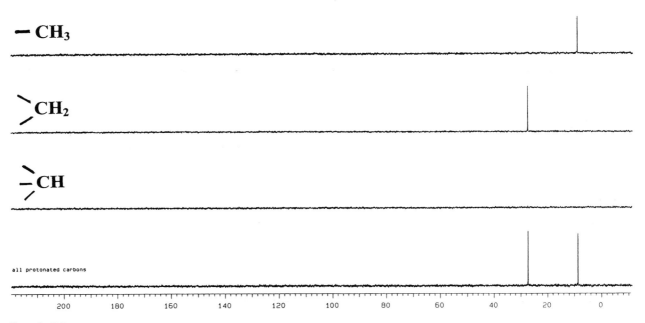

Sample 5.2

DECOUPLED SPECTRUM
C_3H_8O

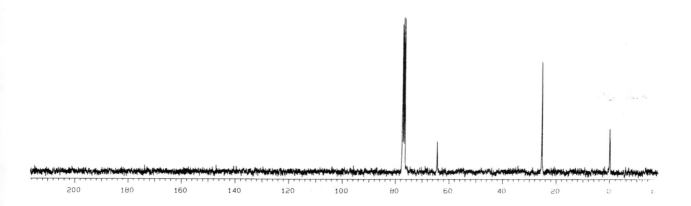

DEPT SPECTRUM

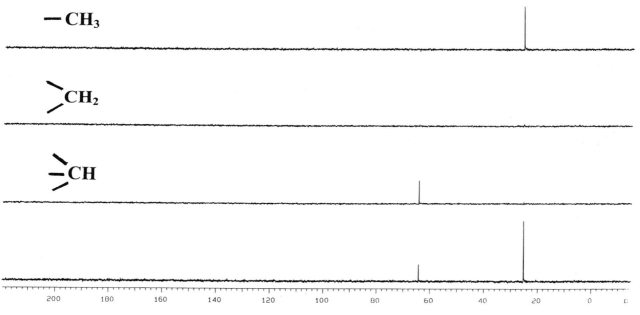

Sample 5.3

DECOUPLED SPECTRUM
C₈H₈O₂

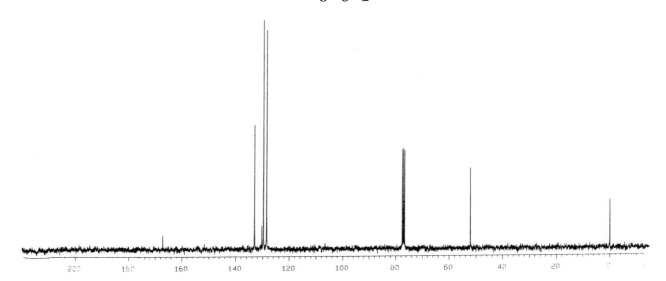

DEPT SPECTRUM

— CH₃

⟩CH₂

⟩CH

all protonated carbons

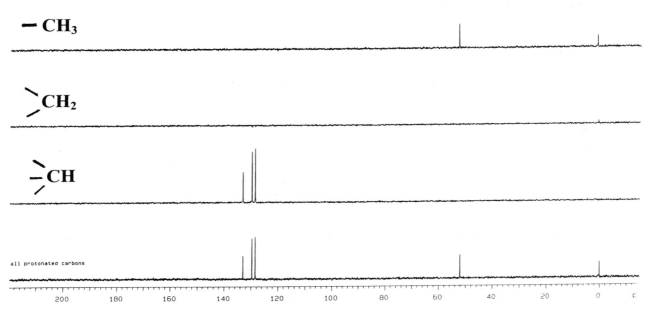

Sample 5.4

DECOUPLED SPECTRUM
C_5H_{12}

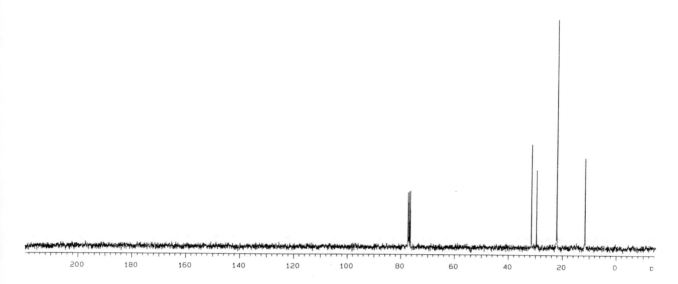

DEPT SPECTRUM

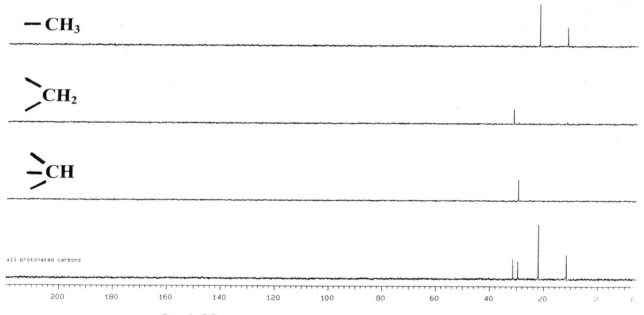

Sample 5.5

DECOUPLED SPECTRUM
C₆H₁₀

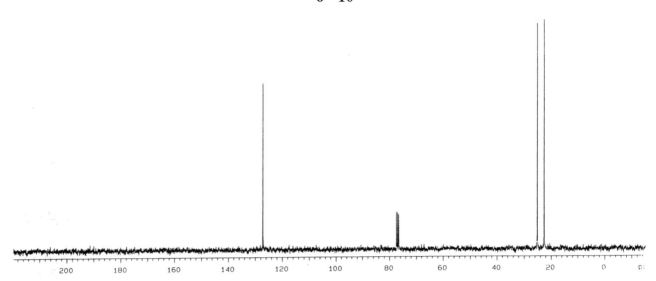

DEPT SPECTRUM

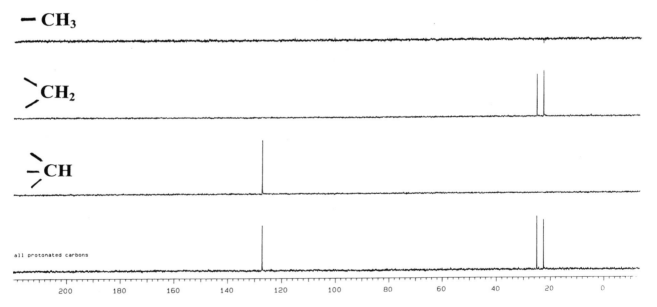

Sample 5.6

DECOUPLED SPECTRUM
$C_8H_{10}O$

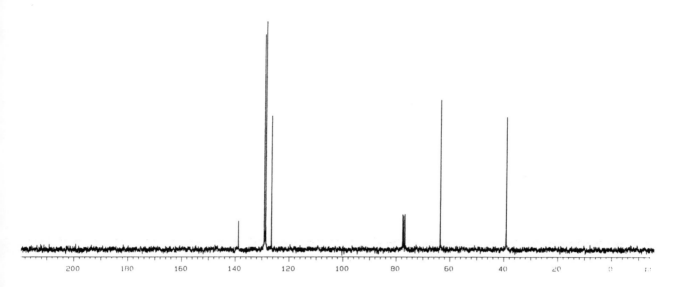

DEPT SPECTRUM

$-CH_3$

$>CH_2$

$>CH$

all protonated carbons

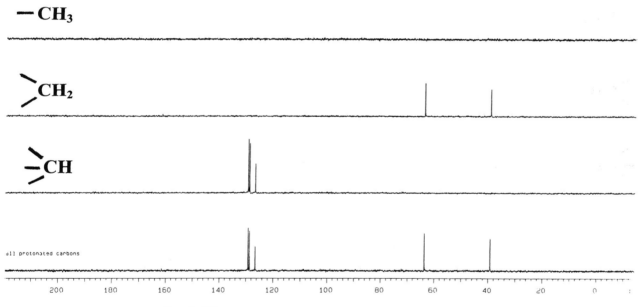

Sample 5.7

DECOUPLED SPECTRUM
C₁₀H₁₅N

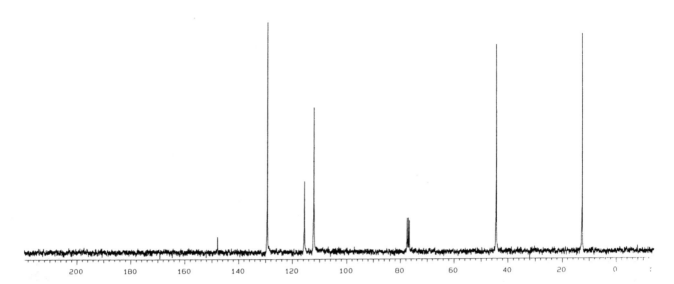

DEPT SPECTRUM

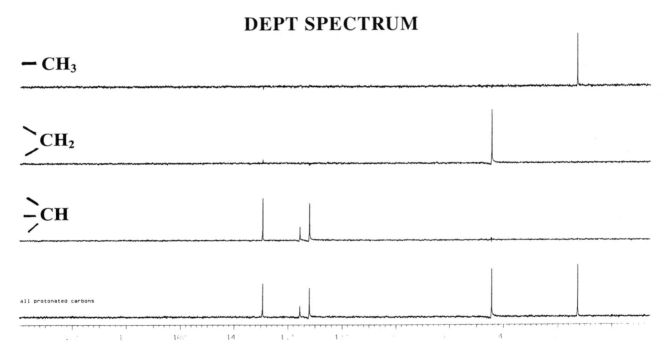

Sample 5.8

DECOUPLED SPECTRUM
$C_8H_{18}O$

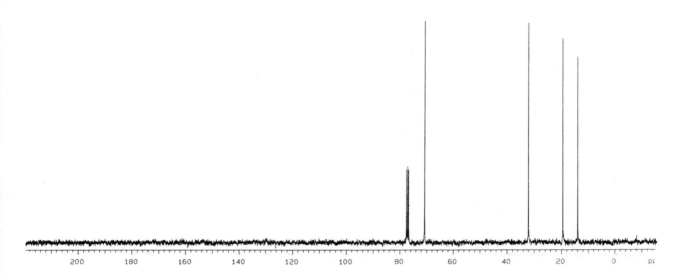

DEPT SPECTRUM

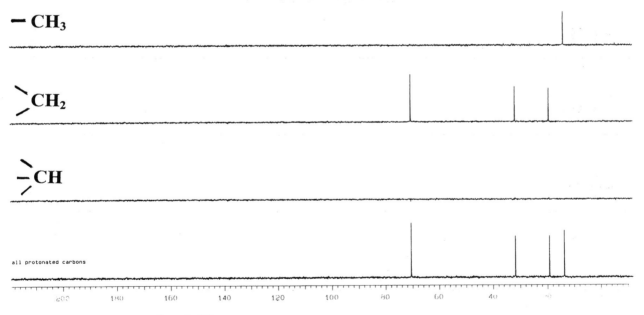

Sample 5.9

DECOUPLED SPECTRUM
C₉H₁₂

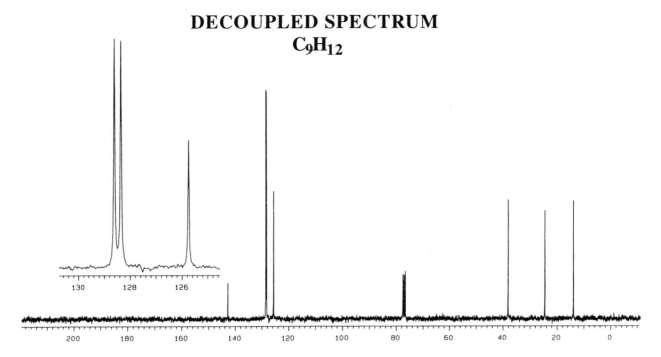

DEPT SPECTRUM

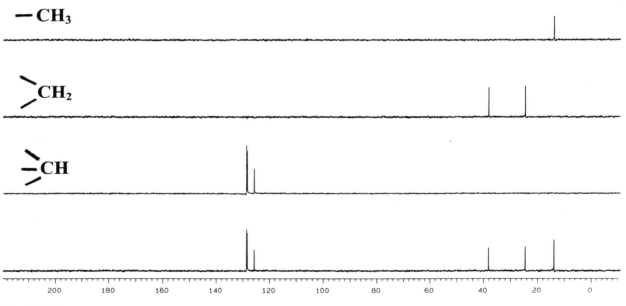

Sample 5.10

CHAPTER 6

Problems in Interpreting Infrared Spectra and ^{13}C Nuclear Magnetic Resonance Spectra

In this chapter, as in all further chapters, only decoupled ^{13}C NMR spectra are given. Instead of DEPT spectra, an indication of whether each peak represents C, CH, CH_2 or CH_3 is given on the decoupled spectrum.

Combining infrared and ^{13}C NMR spectra greatly extends the range of compounds that we can identify. We still need molecular formulae, and these are provided. Although we can identify some functional groups by ^{13}C NMR, infrared is a better technique for all except alkenes and aromatic rings which sometimes have small infrared peaks but are easily seen by ^{13}C NMR. The carbonyl group in particular should always be examined by infrared spectroscopy, and the two techniques supplement each other when identifying alkyne and nitrile groups.

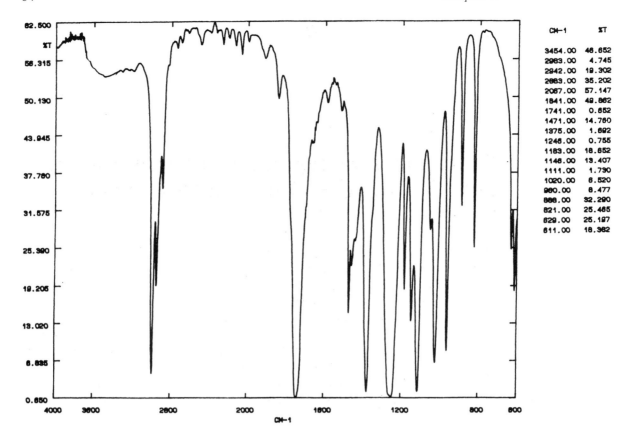

CM-1	%T
3454.00	46.652
2963.00	4.745
2942.00	19.302
2883.00	35.202
2067.00	57.147
1841.00	49.862
1741.00	0.652
1471.00	14.760
1375.00	1.692
1246.00	0.755
1183.00	18.652
1146.00	13.407
1111.00	1.790
1020.00	6.520
980.00	8.477
866.00	32.290
821.00	25.465
629.00	25.197
611.00	18.362

$C_5H_{10}O_2$

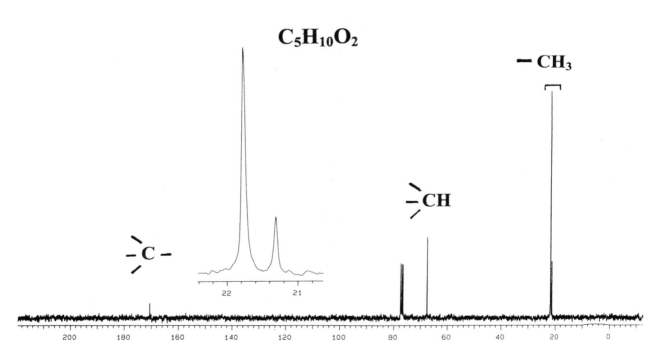

Sample 6.1

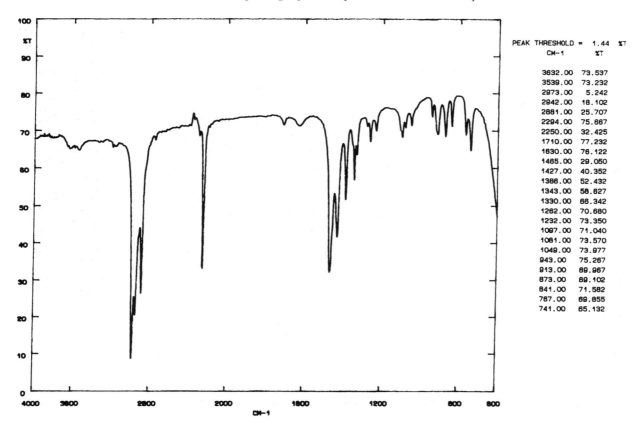

PEAK THRESHOLD = 1.44 %T

CM-1	%T
3632.00	73.537
3539.00	73.232
2973.00	5.242
2942.00	18.102
2881.00	25.707
2294.00	75.667
2250.00	32.425
1710.00	77.232
1630.00	76.122
1465.00	29.050
1427.00	40.352
1386.00	52.432
1343.00	58.627
1330.00	66.342
1262.00	70.680
1232.00	73.350
1097.00	71.040
1061.00	73.570
1049.00	73.977
943.00	75.267
913.00	69.967
873.00	69.102
841.00	71.582
767.00	69.855
741.00	65.132

C_4H_7N

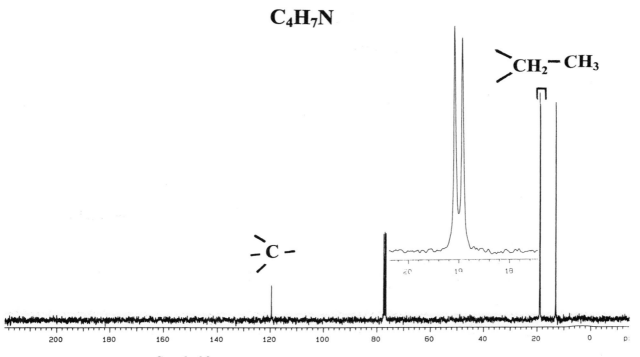

Sample 6.2

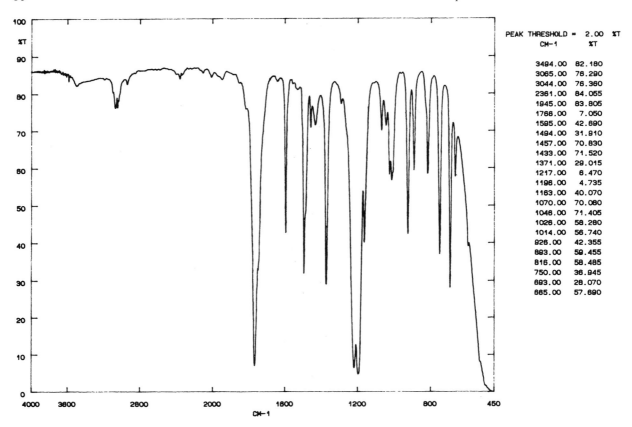

PEAK THRESHOLD = 2.00 %T

CM-1	%T
3494.00	82.180
3065.00	76.290
3044.00	76.360
2361.00	84.055
1945.00	83.805
1768.00	7.050
1595.00	42.690
1494.00	31.910
1457.00	70.830
1433.00	71.520
1371.00	29.015
1217.00	6.470
1196.00	4.735
1163.00	40.070
1070.00	70.080
1046.00	71.405
1026.00	58.280
1014.00	58.740
926.00	42.355
893.00	59.455
816.00	58.485
750.00	36.945
693.00	28.070
665.00	57.690

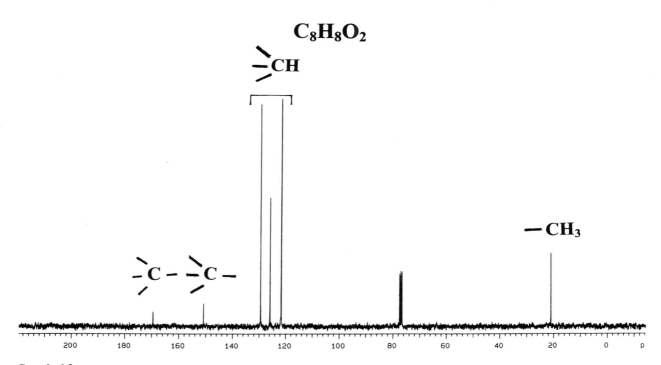

$C_8H_8O_2$

Sample 6.3

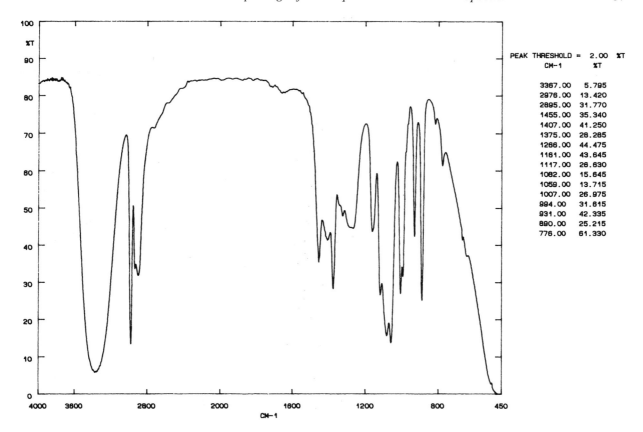

PEAK THRESHOLD = 2.00 %T

CM-1	%T
3367.00	5.795
2976.00	13.420
2895.00	31.770
1455.00	35.340
1407.00	41.250
1375.00	28.285
1266.00	44.475
1161.00	43.645
1117.00	26.630
1082.00	15.645
1059.00	13.715
1007.00	26.975
994.00	31.615
931.00	42.335
890.00	25.215
776.00	61.330

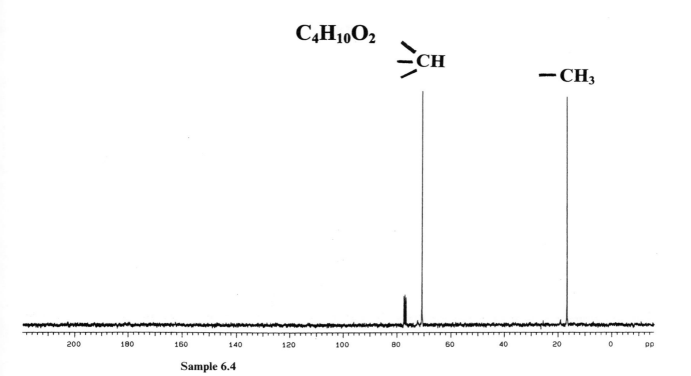

$C_4H_{10}O_2$

Sample 6.4

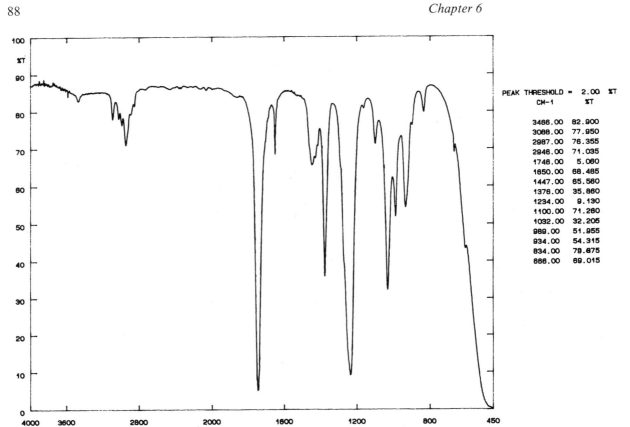

PEAK THRESHOLD = 2.00 %T

CM-1	%T
3466.00	82.900
3088.00	77.950
2987.00	76.355
2946.00	71.035
1746.00	5.060
1650.00	68.485
1447.00	65.560
1376.00	35.860
1234.00	9.130
1100.00	71.260
1032.00	32.205
989.00	51.955
934.00	54.315
834.00	79.875
686.00	69.015

$C_5H_8O_2$

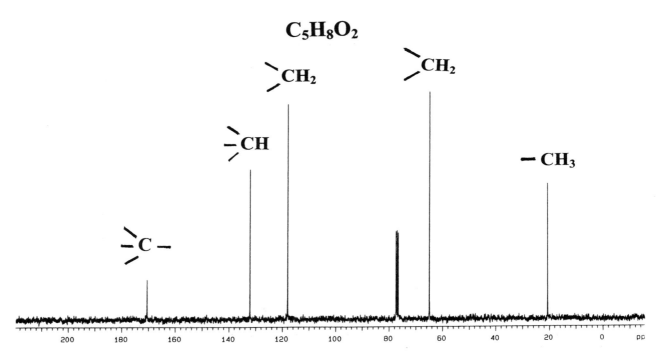

Sample 6.5

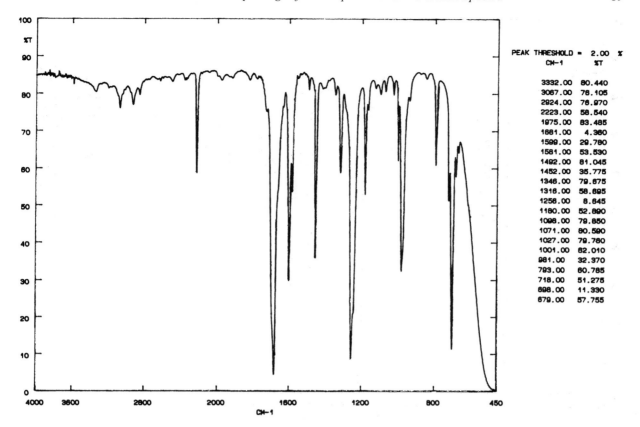

PEAK THRESHOLD = 2.00 %

CM-1	%T
3332.00	80.440
3067.00	78.105
2924.00	78.970
2223.00	58.540
1975.00	83.485
1681.00	4.360
1599.00	29.780
1581.00	53.530
1492.00	81.045
1452.00	35.775
1346.00	79.675
1316.00	58.695
1256.00	8.645
1180.00	52.890
1098.00	79.850
1071.00	80.590
1027.00	79.760
1001.00	82.010
981.00	32.370
793.00	60.785
718.00	51.275
698.00	11.330
679.00	57.755

$$C_8H_5ON$$

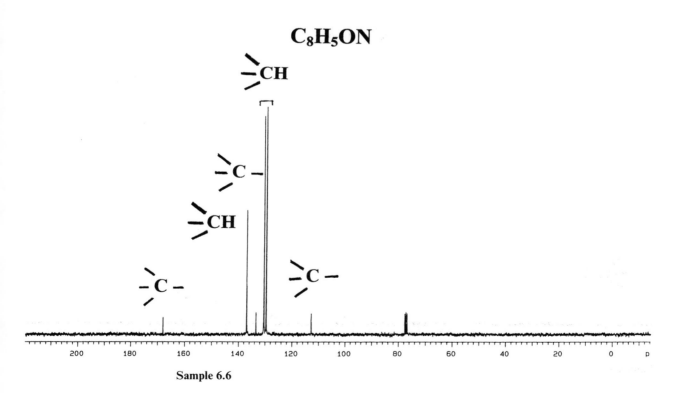

Sample 6.6

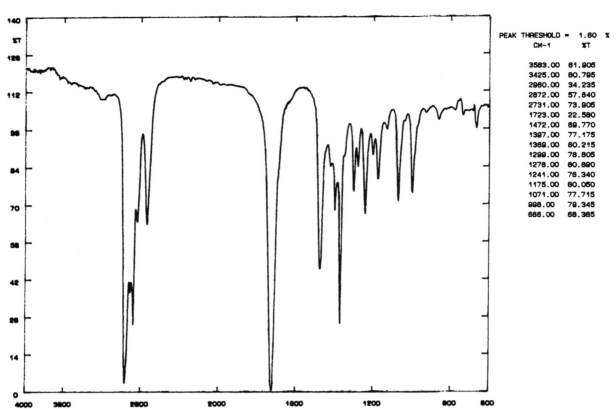

PEAK THRESHOLD = 1.60 %

CM-1	%T
3583.00	81.905
3425.00	80.795
2960.00	34.235
2872.00	57.640
2731.00	73.905
1723.00	22.580
1472.00	69.770
1397.00	77.175
1369.00	60.215
1299.00	78.805
1278.00	80.890
1241.00	76.340
1175.00	80.050
1071.00	77.715
998.00	79.345
666.00	68.385

$$C_6H_{12}O$$

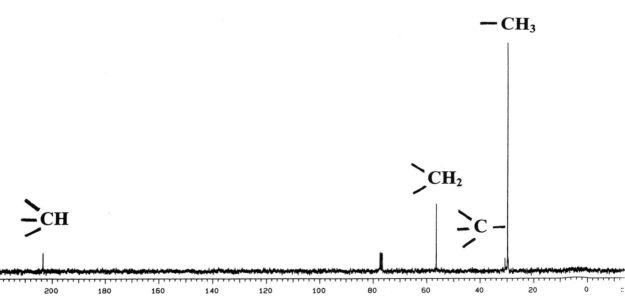

Sample 6.7

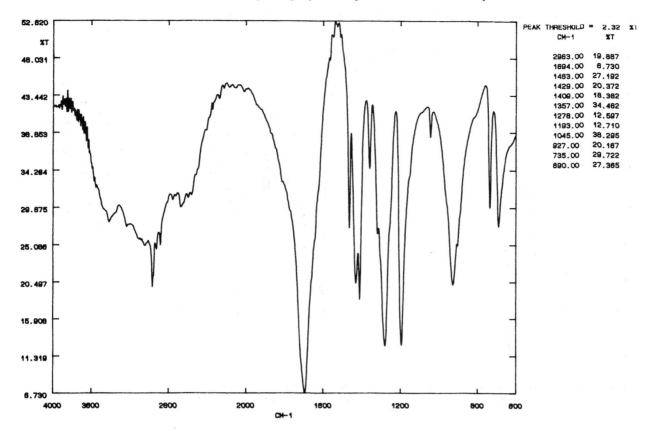

CM-1	%T
2963.00	19.887
1694.00	6.730
1463.00	27.192
1429.00	20.372
1409.00	18.382
1357.00	34.482
1278.00	12.597
1193.00	12.710
1045.00	38.295
927.00	20.167
735.00	29.722
690.00	27.365

PEAK THRESHOLD = 2.32 %T

$C_6H_{10}O_4$

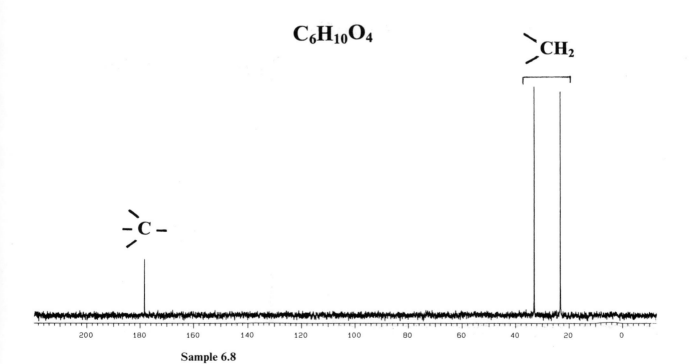

Sample 6.8

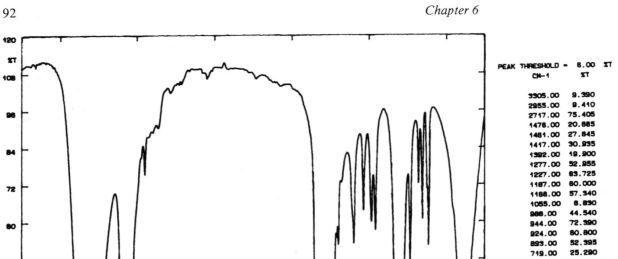

PEAK THRESHOLD =   6.00   %T

CM-1	%T
3305.00	9.390
2955.00	9.410
2717.00	75.405
1476.00	20.665
1461.00	27.645
1417.00	30.935
1392.00	19.900
1277.00	52.955
1227.00	63.725
1187.00	60.000
1186.00	57.340
1055.00	8.630
986.00	44.540
944.00	72.390
924.00	60.800
893.00	52.395
719.00	25.290

$$C_5H_{12}O_2$$

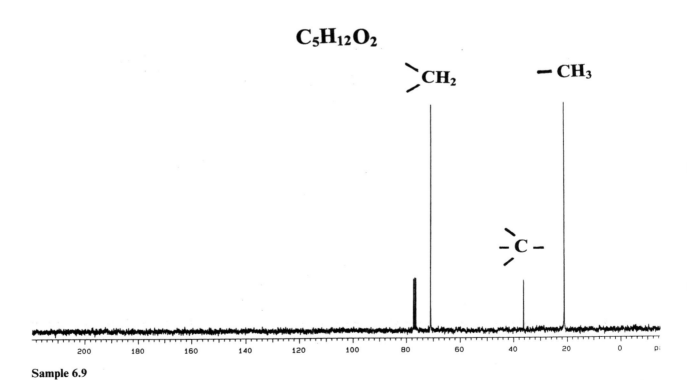

Sample 6.9

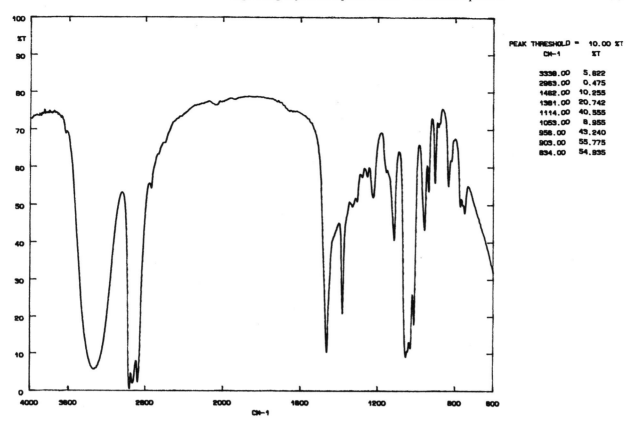

PEAK THRESHOLD = 10.00 %T

CM-1	%T
3338.00	5.822
2963.00	0.475
1462.00	10.255
1381.00	20.742
1114.00	40.555
1053.00	8.955
956.00	43.240
903.00	55.775
834.00	54.935

$C_6H_{14}O$

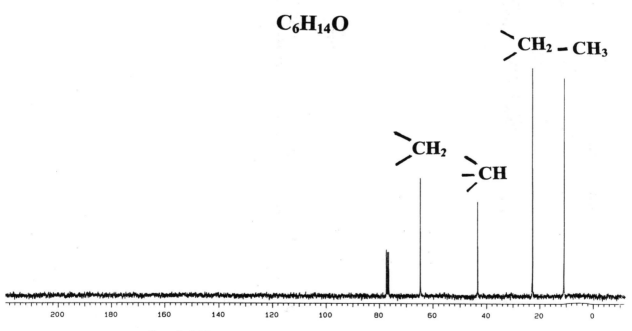

Sample 6.10

CHAPTER 7

Problems in Interpreting Mass Spectra and ^{13}C Nuclear Magnetic Resonance Spectra

In this chapter, we combine mass spectra with ^{13}C NMR spectra, so we no longer need molecular formulae. The techniques again complement each other; minimally, mass spectra give us the molecular weight and the halogen content, while ^{13}C NMR gives us carbonyl groups, double bonds, aromatic rings, alkyne and nitrile groups. Both techniques can contribute a lot more than this to a structure. Mass spectra can tell us of the presence of nitrogen (tertiary amines are otherwise very difficult to detect) and a McLafferty rearrangement can give complete identification by itself in a few cases. The ^{13}C NMR spectrum can identify individual carbon atoms, and mass spectra may then be able to say if a CH_3 is part of an ethyl, propyl or butyl group. The method is particularly useful for identifying symmetrical and cyclic structures.

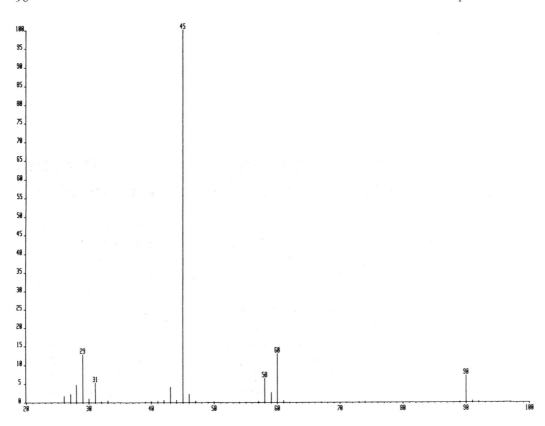

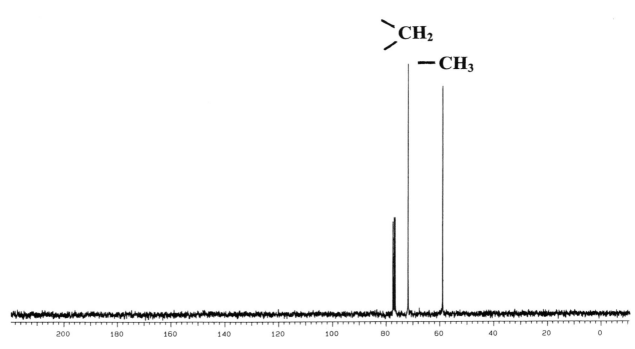

Sample 7.1

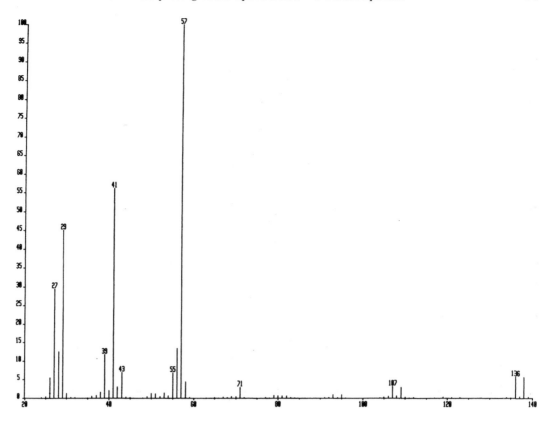

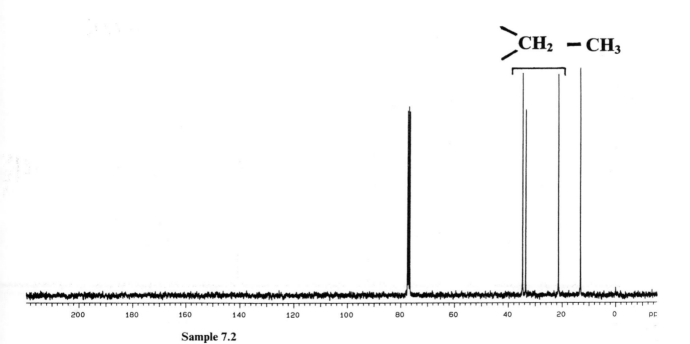

Sample 7.2

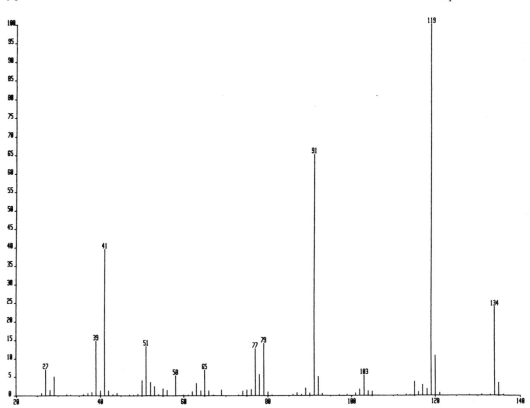

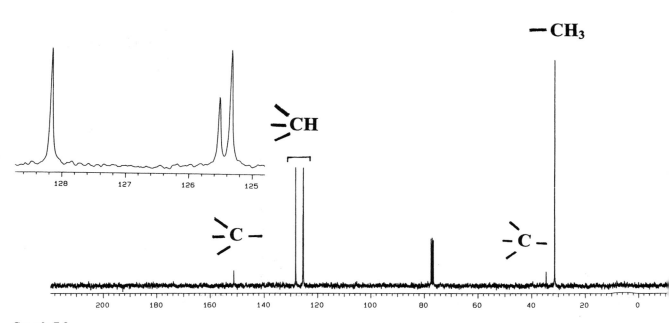

Sample 7.3

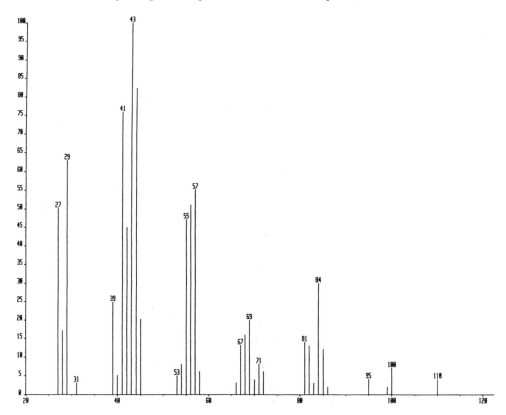

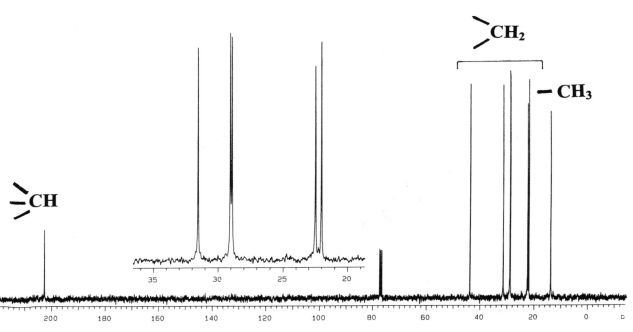

Sample 7.4

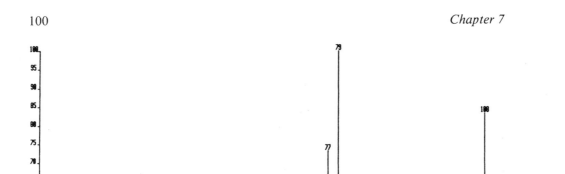

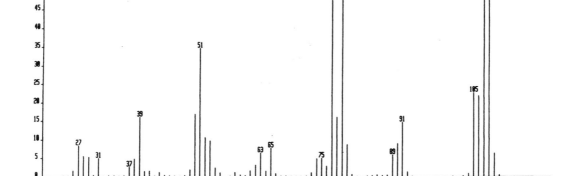

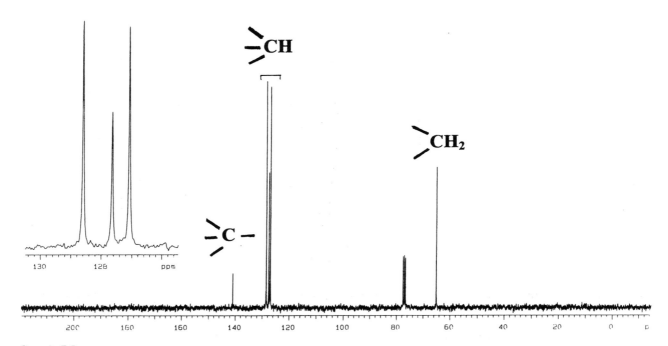

Sample 7.5

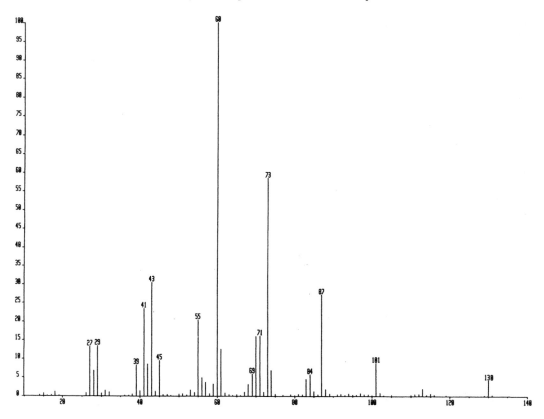

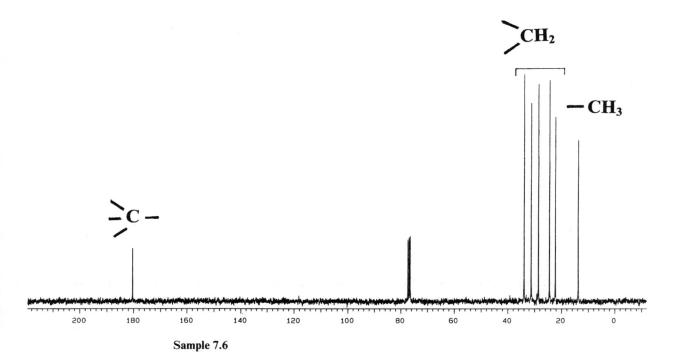

Sample 7.6

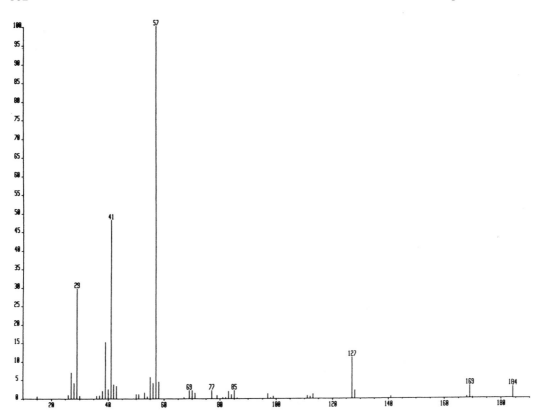

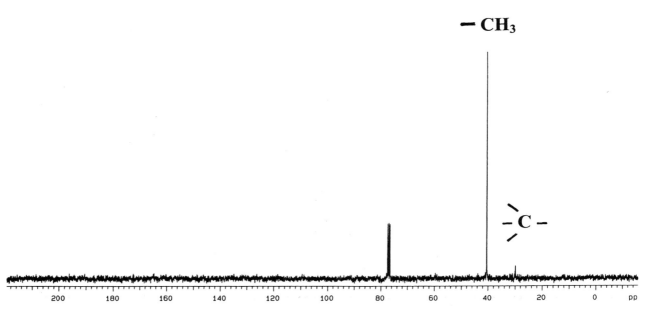

Sample 7.7

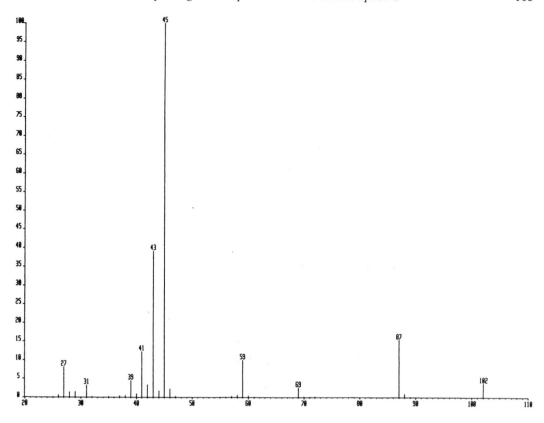

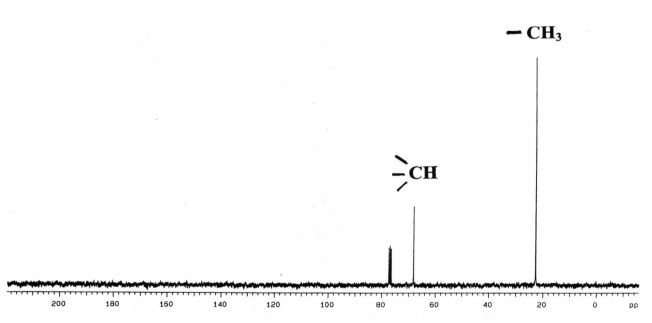

Sample 7.8

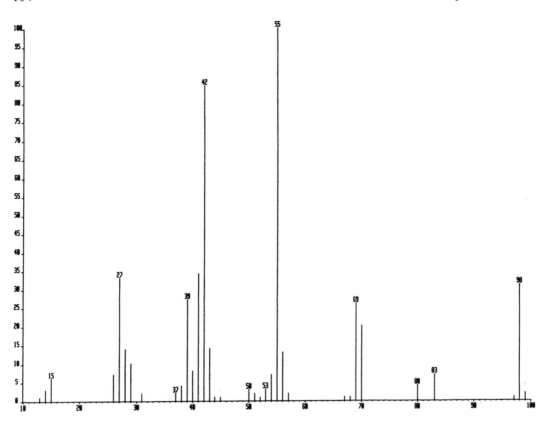

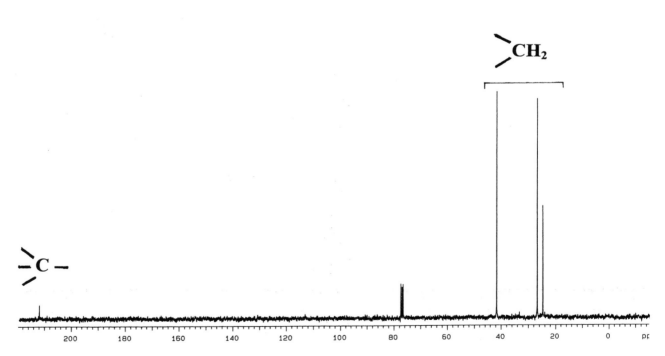

Sample 7.9

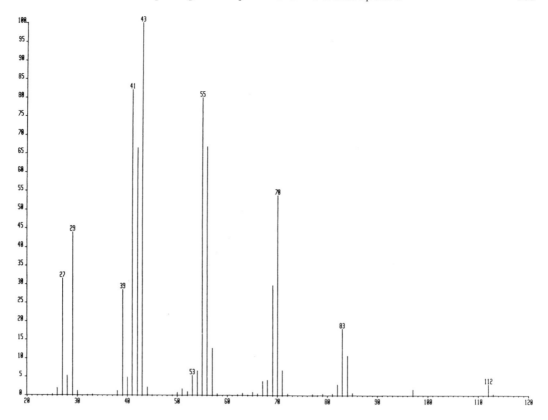

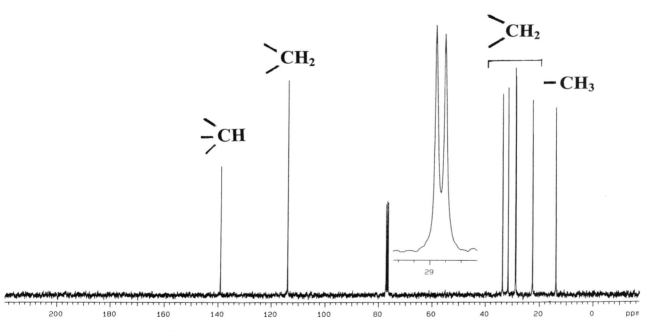

Sample 7.10

CHAPTER 8

Problems in Interpreting Infrared Spectra, Mass Spectra, Ultraviolet Spectra and ^{13}C Nuclear Magnetic Resonance Spectra

We now combine four techniques, which enable us to get a full analysis of possible functional groups, give us the molecular weight and a lot of structural information about our samples. We are still restricted by branched chain compounds, and in some cases cannot distinguish between isomers, but these techniques do well on cyclic systems.

When combining these techniques it is usual to start with infrared, in order to identify the functional groups. Next, study the mass spectrum to obtain the molecular weight, halogen content and main splittings. The ultraviolet spectrum tells us about any conjugation. The ^{13}C NMR spectrum should be checked to see if the C and H content and the functional groups add up to the molecular weight. If not, then the molecule has sufficient symmetry to place some atoms in identical environments.

Always remember, when you have assigned a structure, to check that your assignment is consistent with all the data from all the techniques.

All the samples in the exercise can be identified using the data provided.

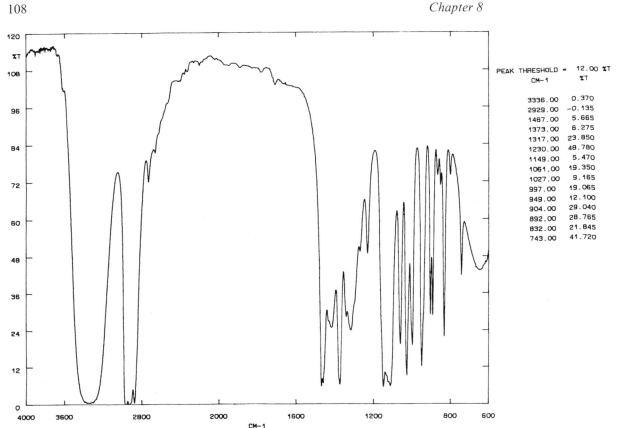

PEAK THRESHOLD = 12.00 %T

CM-1	%T
3336.00	0.370
2929.00	-0.135
1467.00	5.665
1373.00	6.275
1317.00	23.850
1230.00	48.780
1149.00	5.470
1061.00	19.350
1027.00	9.165
997.00	19.065
949.00	12.100
904.00	29.040
892.00	28.765
832.00	21.845
743.00	41.720

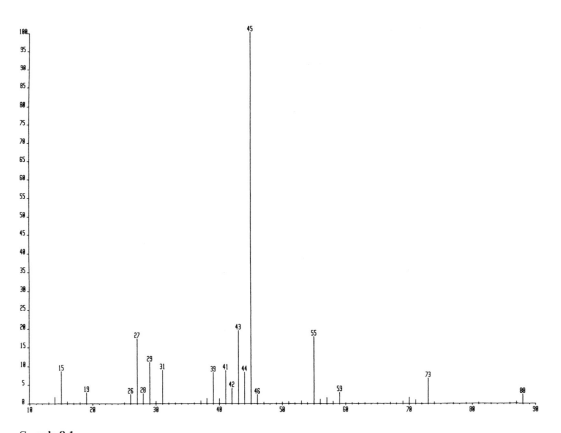

Sample 8.1

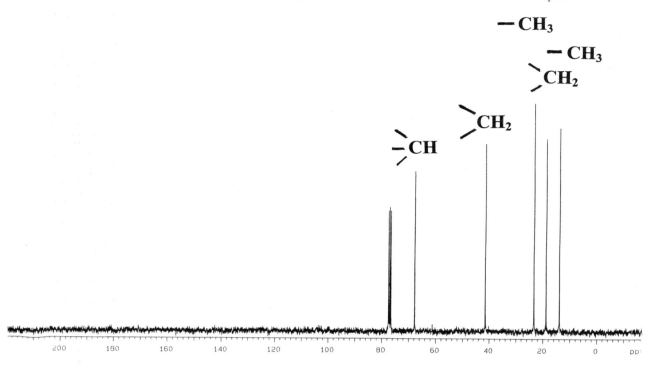

Sample 8.1 *Continued*

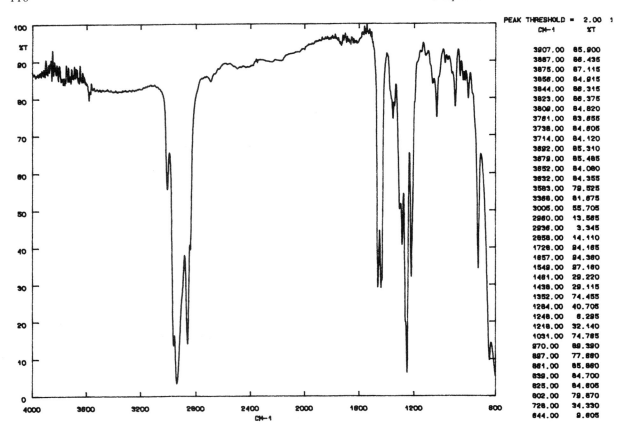

PEAK THRESHOLD = 2.00

CM-1	%T
3907.00	85.900
3887.00	86.435
3875.00	87.115
3856.00	84.915
3844.00	86.315
3823.00	86.375
3809.00	84.820
3781.00	83.855
3738.00	84.605
3714.00	84.120
3692.00	85.310
3679.00	85.485
3652.00	84.080
3632.00	84.355
3583.00	79.525
3366.00	81.675
3005.00	55.705
2960.00	13.585
2936.00	3.345
2858.00	14.110
1726.00	94.165
1657.00	94.360
1549.00	97.160
1461.00	29.220
1438.00	29.115
1352.00	74.455
1284.00	40.705
1246.00	6.295
1218.00	32.140
1031.00	74.785
970.00	89.390
897.00	77.860
861.00	85.860
839.00	84.700
825.00	84.605
802.00	79.870
728.00	34.330
644.00	9.605

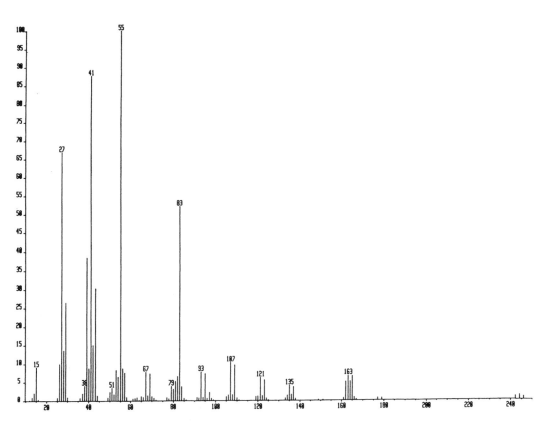

Sample 8.2

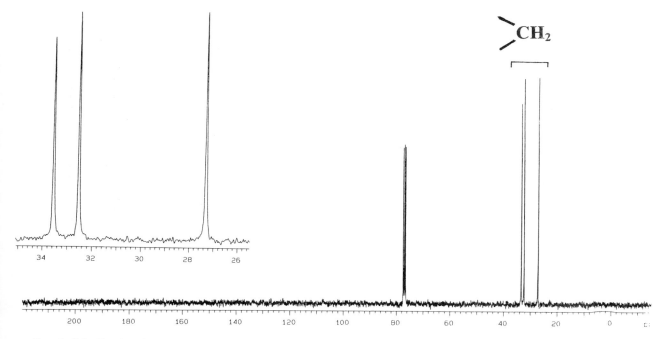

Sample 8.2 *Continued*

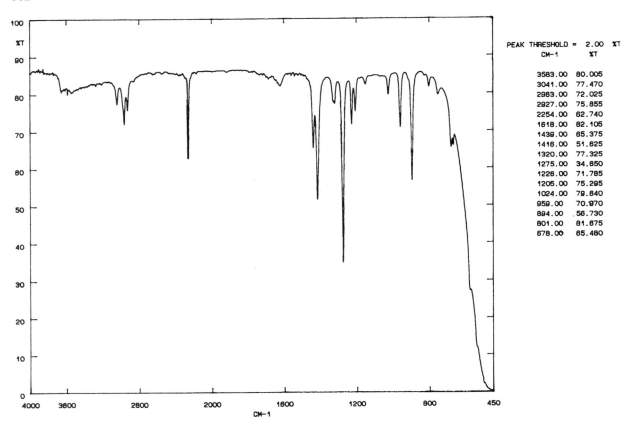

PEAK THRESHOLD = 2.00 %T

CM-1	%T
3583.00	80.005
3041.00	77.470
2963.00	72.025
2927.00	75.855
2254.00	62.740
1618.00	82.105
1439.00	65.375
1416.00	51.625
1320.00	77.325
1275.00	34.650
1226.00	71.785
1205.00	75.295
1024.00	79.640
959.00	70.970
894.00	56.730
801.00	81.675
678.00	65.480

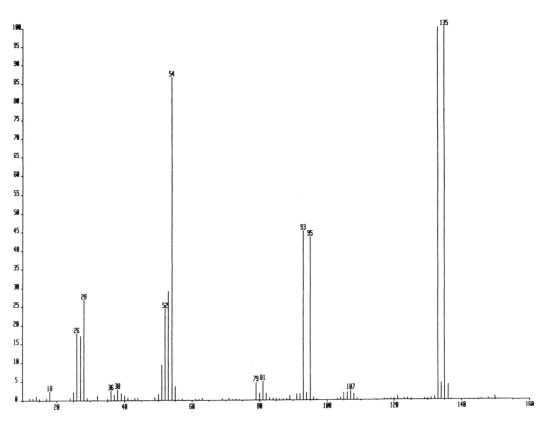

Sample 8.3

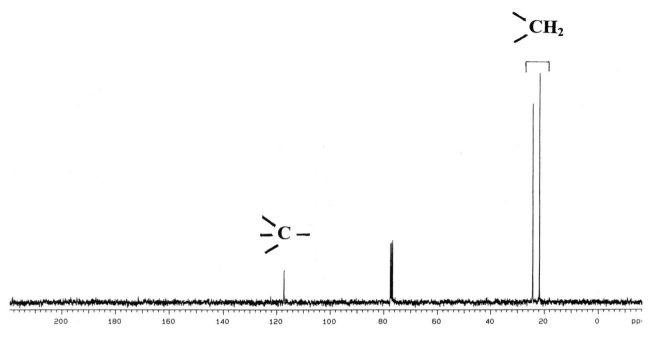

Sample 8.3 *Continued*

PEAK THRESHOLD = 1.07 1
CH-1   %T

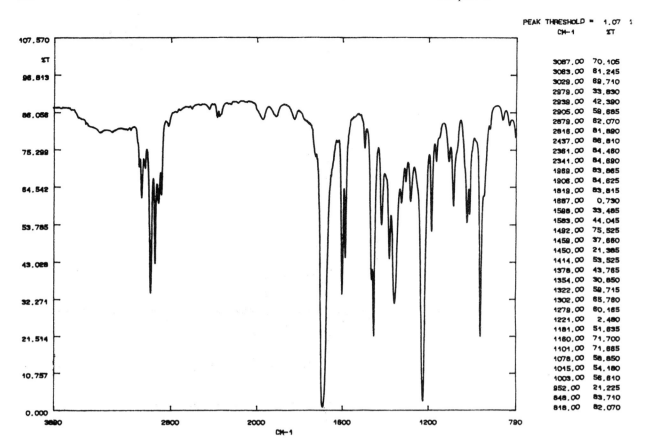

CH-1	%T
3067.00	70.105
3053.00	61.245
3029.00	69.710
2979.00	33.830
2939.00	42.390
2905.00	59.685
2879.00	82.070
2816.00	81.690
2437.00	86.810
2361.00	84.460
2341.00	84.690
1969.00	83.865
1906.00	84.625
1819.00	83.815
1687.00	0.730
1596.00	33.485
1583.00	44.045
1492.00	75.525
1459.00	37.660
1450.00	21.385
1414.00	53.525
1378.00	43.785
1354.00	30.850
1322.00	59.715
1302.00	65.780
1279.00	60.165
1221.00	2.480
1181.00	51.635
1160.00	71.700
1101.00	71.685
1078.00	58.850
1015.00	54.180
1003.00	56.610
952.00	21.225
848.00	83.710
818.00	82.070

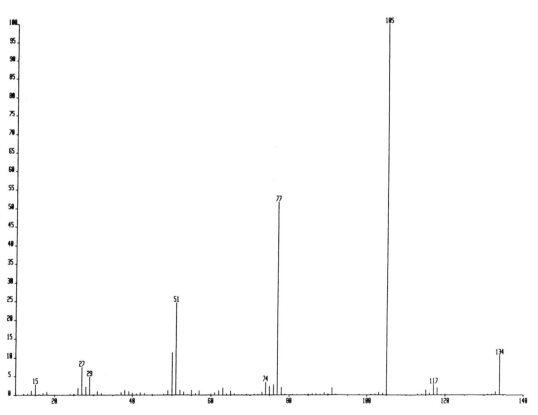

Sample 8.4

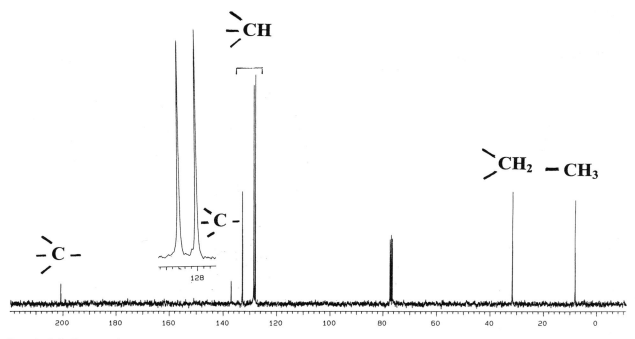

Sample 8.4 *Continued*

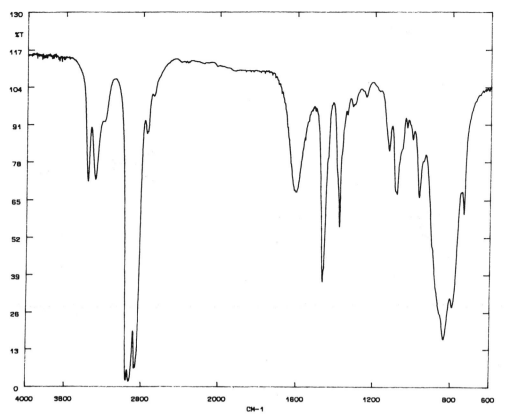

PEAK THRESHOLD = 13.00 %T

CM-1	%T
3367.00	71.300
3290.00	71.900
2927.00	1.785
2873.00	6.305
1607.00	67.700
1465.00	36.655
1378.00	55.770
1082.00	67.030
967.00	65.950
836.00	16.500

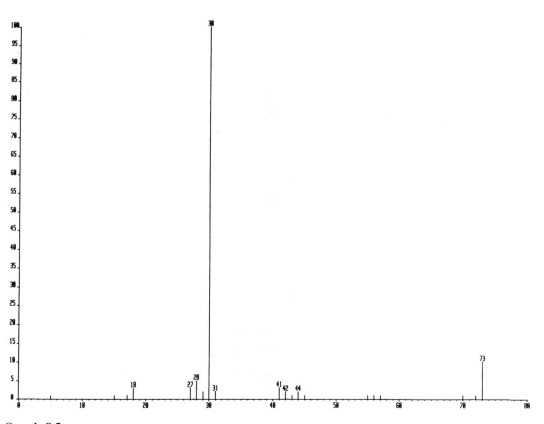

Sample 8.5

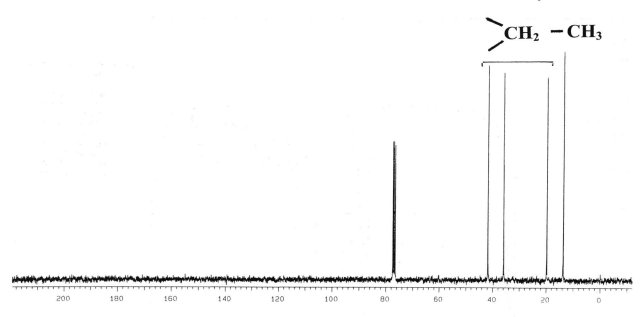

Sample 8.5 *Continued*

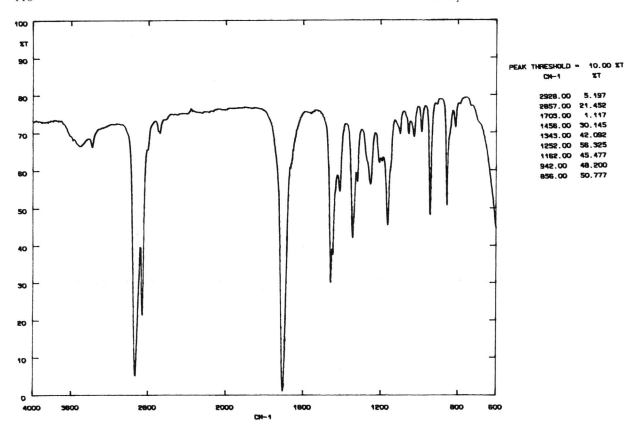

PEAK THRESHOLD = 10.00 %T

CM-1	%T
2928.00	5.197
2857.00	21.452
1703.00	1.117
1458.00	30.145
1343.00	42.092
1252.00	58.325
1162.00	45.477
942.00	48.200
856.00	50.777

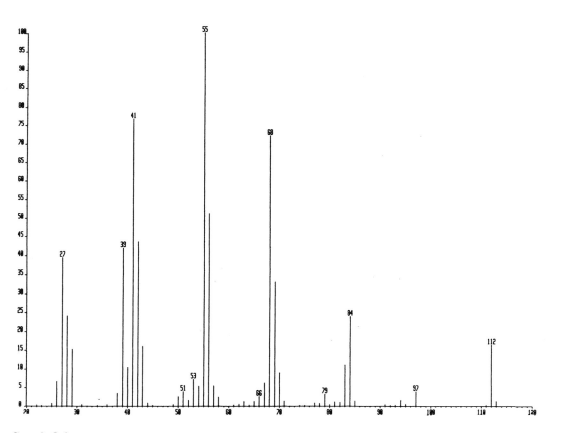

Sample 8.6

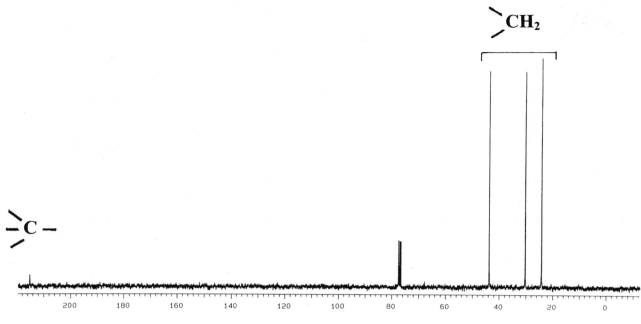

Sample 8.6 *Continued*

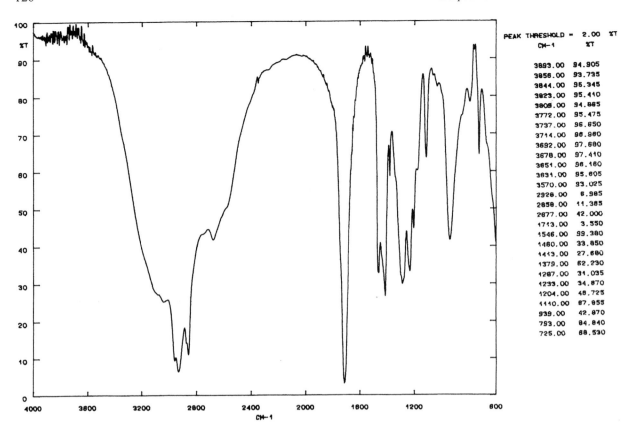

PEAK THRESHOLD =   2.00 %T
CM-1	%T
3893.00	94.905
3856.00	93.735
3844.00	95.345
3823.00	95.410
3805.00	94.885
3772.00	95.475
3737.00	96.850
3714.00	96.980
3692.00	97.680
3678.00	97.410
3651.00	96.160
3631.00	95.605
3570.00	93.025
2926.00	6.985
2858.00	11.385
2677.00	42.000
1713.00	3.550
1546.00	99.380
1460.00	33.850
1443.00	27.680
1379.00	62.230
1287.00	31.035
1233.00	34.870
1204.00	48.725
1110.00	87.955
939.00	42.870
793.00	84.840
725.00	68.530

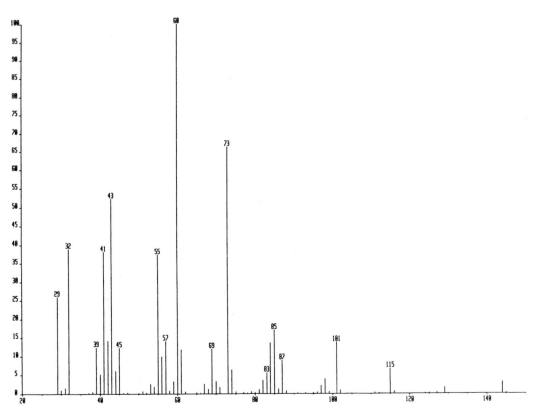

Sample 8.7

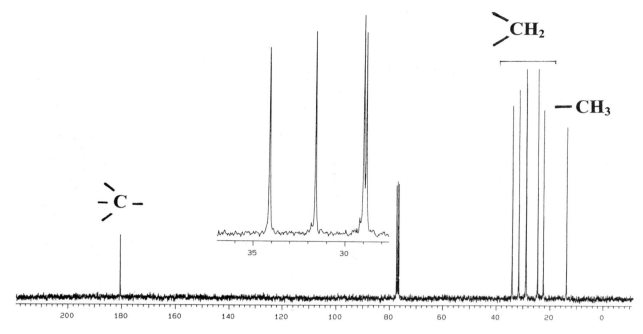

Sample 8.7 *Continued*

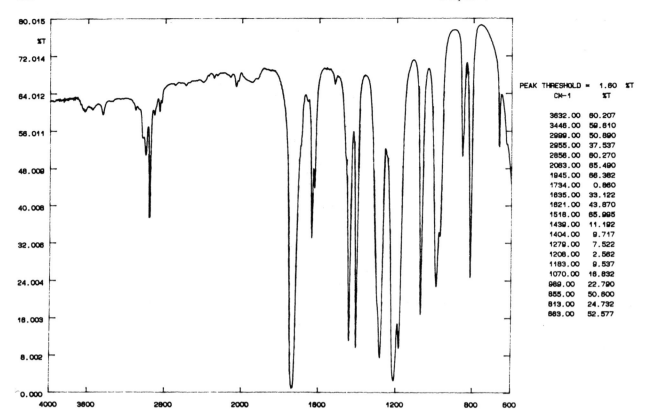

PEAK THRESHOLD =   1.60 %T

CM-1	%T
3632.00	60.207
3446.00	59.610
2989.00	50.890
2955.00	37.537
2856.00	60.270
2063.00	65.490
1945.00	66.382
1734.00	0.860
1635.00	33.122
1621.00	43.870
1518.00	65.995
1439.00	11.192
1404.00	9.717
1279.00	7.522
1208.00	2.582
1183.00	9.537
1070.00	16.832
969.00	22.790
855.00	50.600
813.00	24.732
663.00	52.577

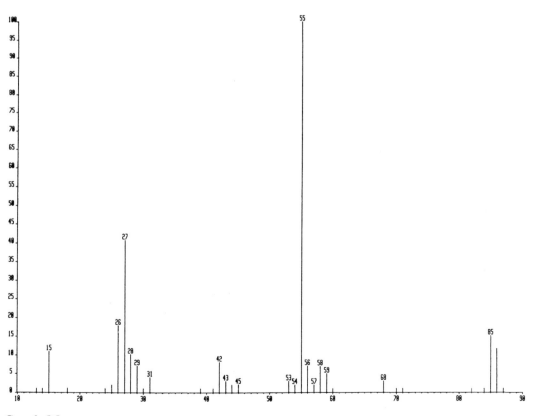

Sample 8.8

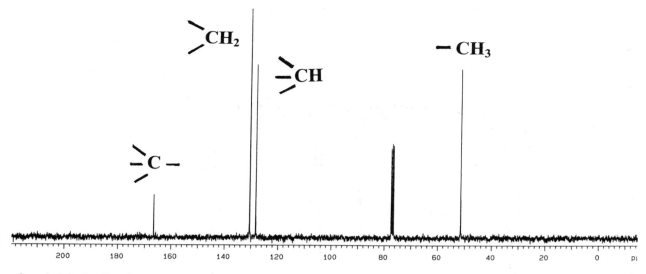

Sample 8.8 *Continued*

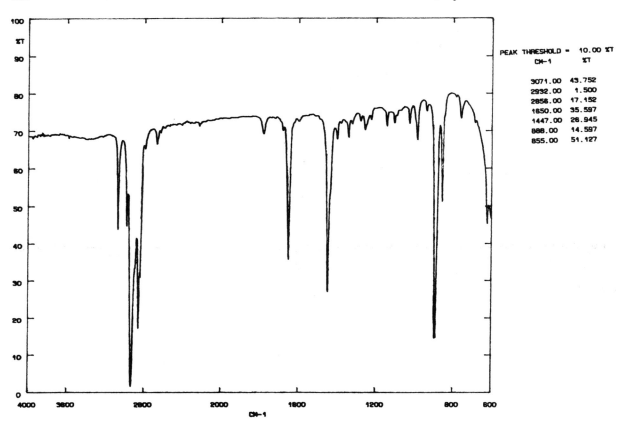

PEAK THRESHOLD =   10.00 %T
      CM-1   %T

    3071.00   43.752
    2932.00   1.500
    2856.00   17.152
    1650.00   35.597
    1447.00   26.945
     888.00   14.597
     855.00   51.127

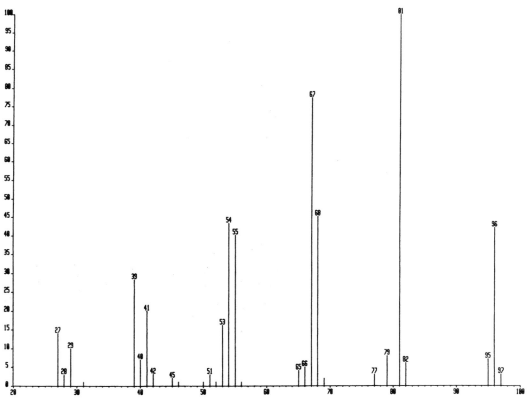

Sample 8.9

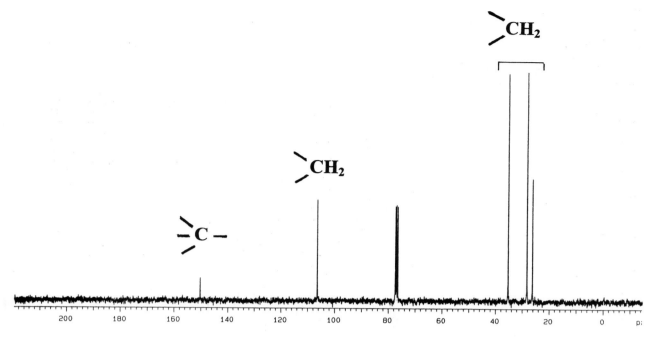

Sample 8.9 *Continued*

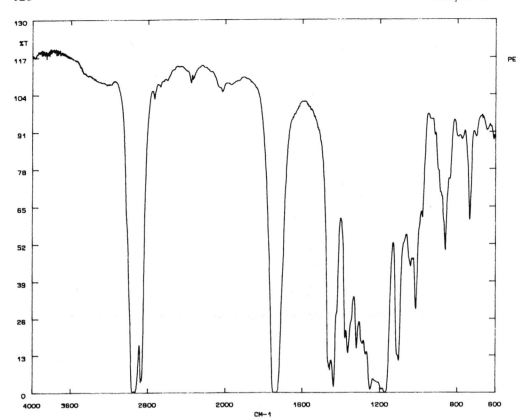

PEAK THRESHOLD =   13.00 %T
        CM-1   %T

        2961.00   -0.015
        2874.00   3.615
        1735.00   -0.360
        1437.00   2.020
        1364.00   13.850
        1173.00   0.030
        1103.00   11.170
        1014.00   29.440
         862.00   49.725
         735.00   60.290

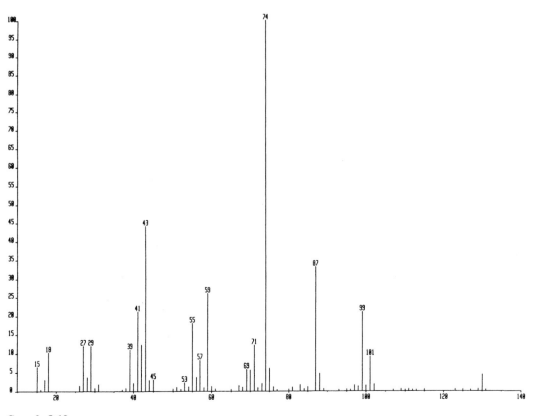

Sample 8.10

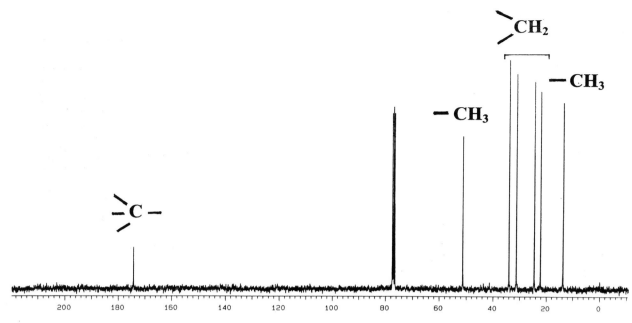

Sample 8.10 *Continued*

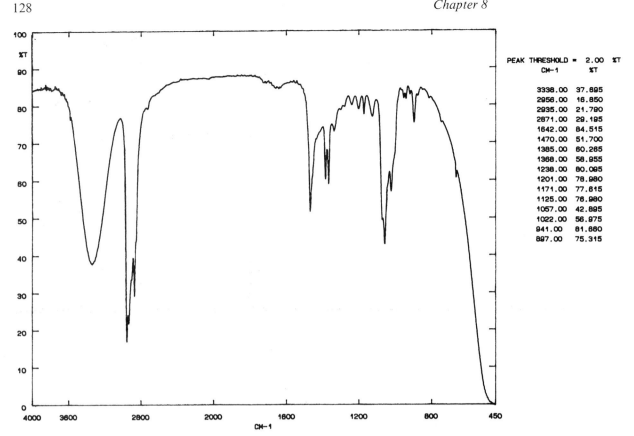

PEAK THRESHOLD = 2.00 %T

CM-1	%T
3338.00	37.895
2956.00	16.850
2935.00	21.790
2871.00	29.195
1642.00	84.515
1470.00	51.700
1385.00	60.265
1368.00	58.955
1238.00	80.095
1201.00	78.980
1171.00	77.615
1125.00	76.980
1057.00	42.895
1022.00	56.975
941.00	81.660
897.00	75.315

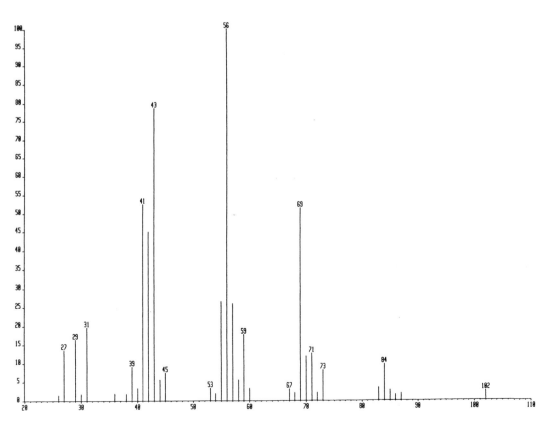

Sample 8.11

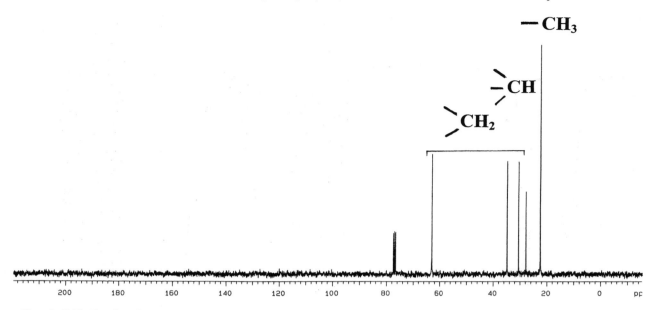

Sample 8.11 *Continued*

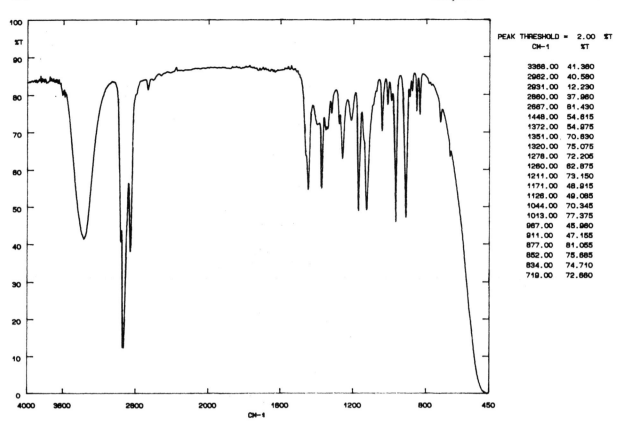

PEAK THRESHOLD = 2.00 %T

CM-1	%T
3368.00	41.360
2962.00	40.580
2931.00	12.230
2860.00	37.960
2667.00	81.430
1448.00	54.615
1372.00	54.975
1351.00	70.630
1320.00	75.075
1278.00	72.205
1260.00	62.875
1211.00	73.150
1171.00	48.915
1126.00	49.085
1044.00	70.345
1013.00	77.375
967.00	45.960
911.00	47.155
877.00	81.055
852.00	75.685
834.00	74.710
719.00	72.660

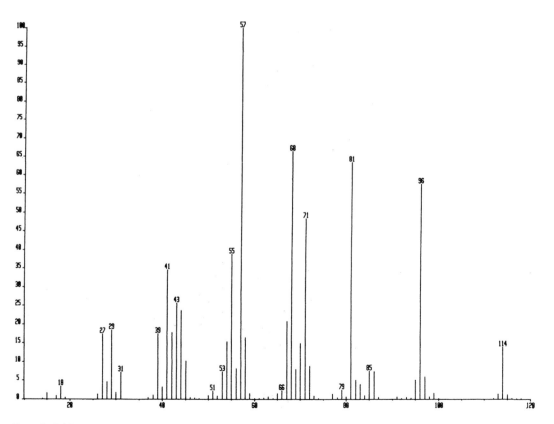

Sample 8.12

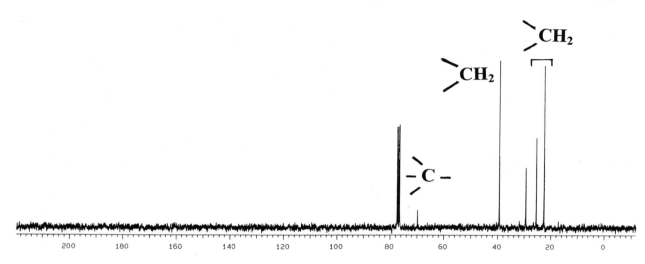

Sample 8.12 *Continued*

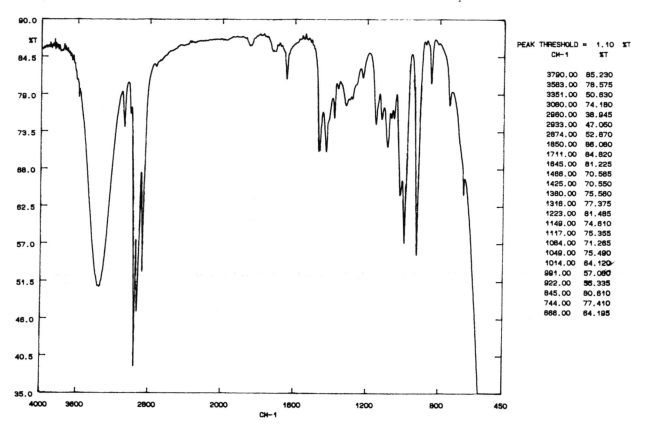

PEAK THRESHOLD = 1.10 %T

CM-1	%T
3790.00	85.230
3583.00	78.575
3351.00	50.630
3080.00	74.180
2960.00	38.945
2933.00	47.050
2874.00	52.870
1850.00	86.080
1711.00	84.820
1645.00	81.225
1466.00	70.565
1425.00	70.550
1380.00	75.560
1316.00	77.375
1223.00	81.485
1149.00	74.610
1117.00	75.355
1084.00	71.265
1049.00	75.490
1014.00	64.120
991.00	57.060
922.00	55.335
845.00	80.610
744.00	77.410
666.00	64.195

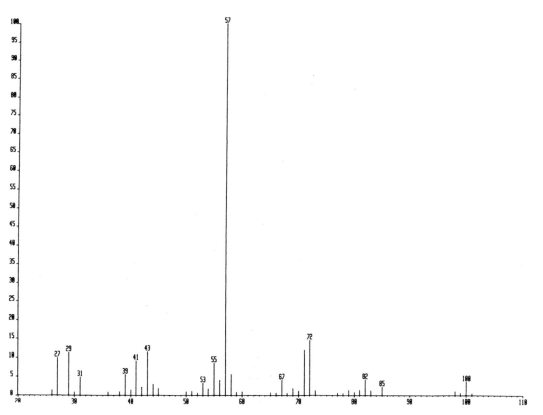

Sample 8.13

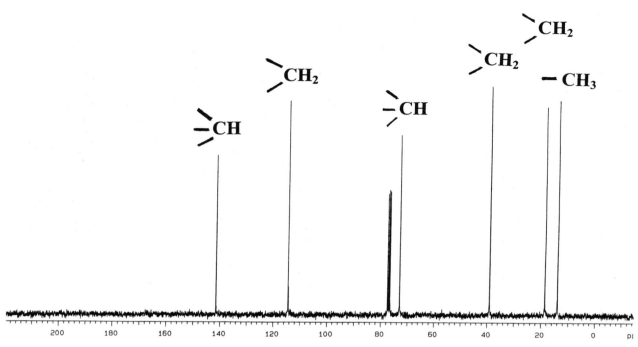

Sample 8.13 *Continued*

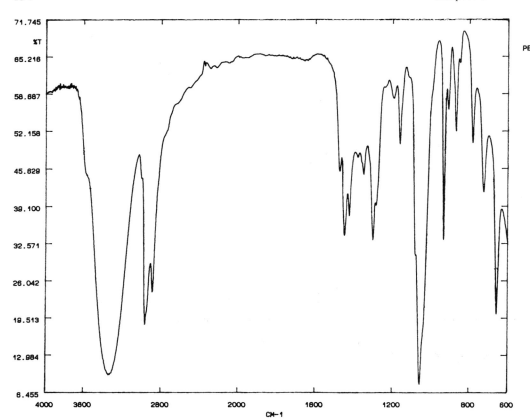

PEAK THRESHOLD =   1.30 %T
     CM-1   %T

    3338.00   9.515
    2963.00   18.347
    2887.00   24.107
    2292.00   63.135
    1473.00   45.172
    1447.00   34.010
    1422.00   37.485
    1348.00   44.620
    1299.00   33.187
    1192.00   57.892
    1161.00   49.840
    1053.00   7.872
     932.00   33.230
     909.00   55.942
     869.00   52.092
     782.00   50.072
     724.00   41.560
     656.00   20.195

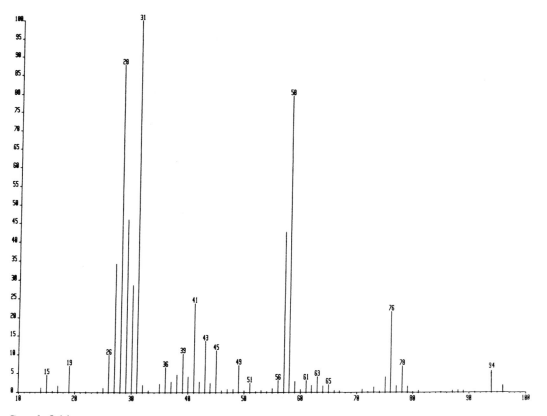

Sample 8.14

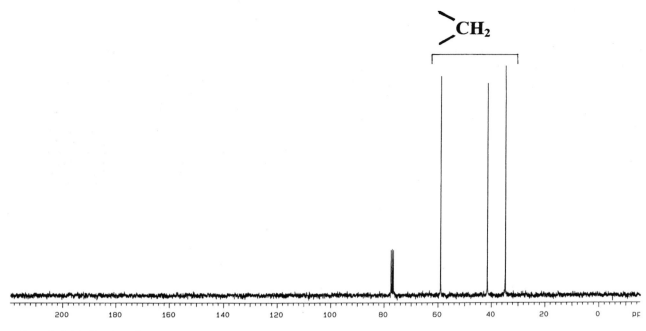

Sample 8.14 *Continued*

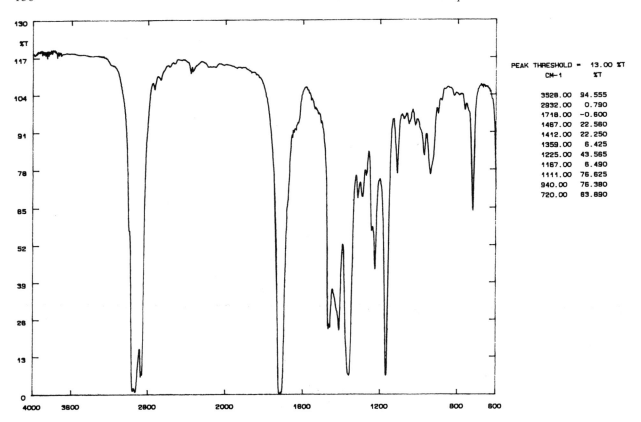

PEAK THRESHOLD =   13.00 %T
     CM-1   %T

    3528.00   94.555
    2932.00   0.790
    1718.00   -0.600
    1467.00   22.560
    1412.00   22.250
    1359.00   6.425
    1225.00   43.565
    1167.00   6.490
    1111.00   76.625
     940.00   76.380
     720.00   63.890

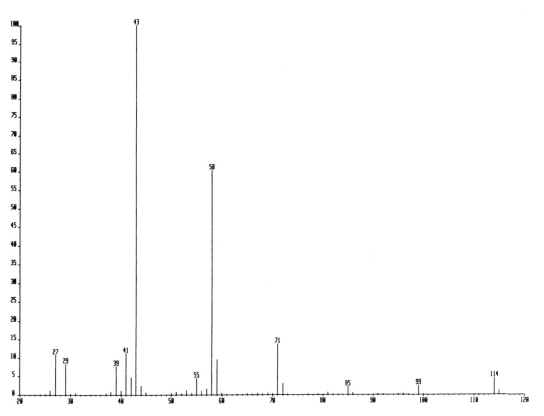

Sample 8.15

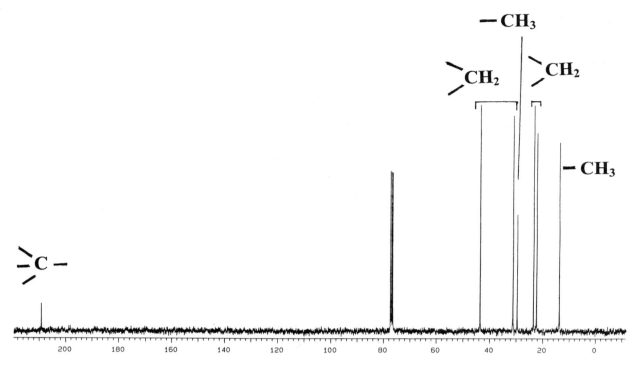

Sample 8.15 *Continued*

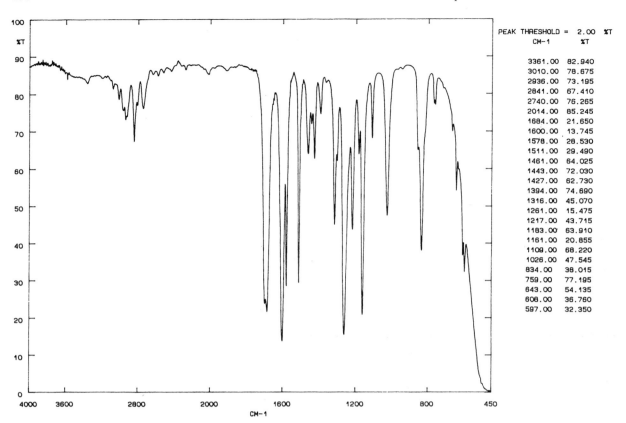

PEAK THRESHOLD = 2.00 %T

CM-1	%T
3361.00	82.940
3010.00	78.675
2936.00	73.195
2841.00	67.410
2740.00	76.265
2014.00	85.245
1684.00	21.650
1600.00	13.745
1578.00	28.530
1511.00	29.490
1461.00	64.025
1443.00	72.030
1427.00	62.730
1394.00	74.690
1316.00	45.070
1261.00	15.475
1217.00	43.715
1183.00	63.910
1161.00	20.855
1109.00	68.220
1026.00	47.545
834.00	38.015
759.00	77.195
643.00	54.135
608.00	36.760
597.00	32.350

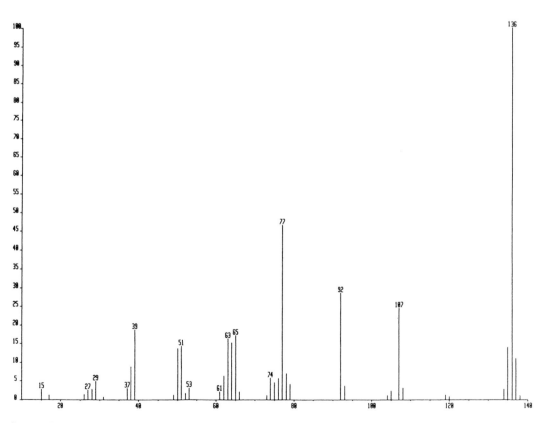

Sample 8.16

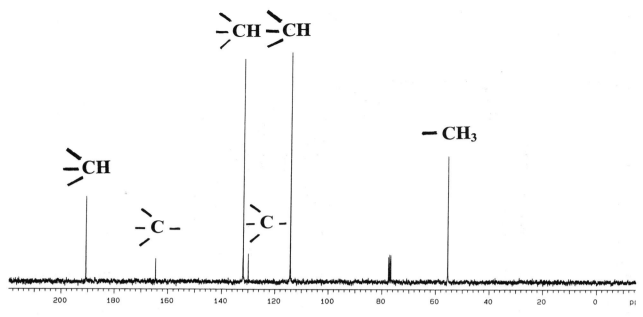

Sample 8.16 *Continued*

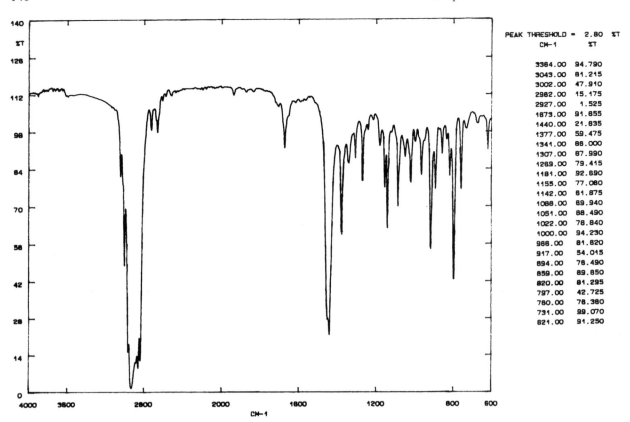

PEAK THRESHOLD =   2.80 %T
       CM-1   %T

     3364.00   94.790
     3043.00   81.215
     3002.00   47.910
     2982.00   15.175
     2927.00   1.525
     1673.00   91.855
     1440.00   21.635
     1377.00   59.475
     1341.00   88.000
     1307.00   87.990
     1289.00   79.415
     1181.00   92.890
     1155.00   77.060
     1142.00   61.875
     1088.00   69.940
     1051.00   88.490
     1022.00   78.840
     1000.00   94.230
      988.00   81.820
      917.00   54.015
      894.00   78.490
      859.00   89.850
      820.00   81.295
      797.00   42.725
      780.00   78.380
      731.00   99.070
      621.00   91.250

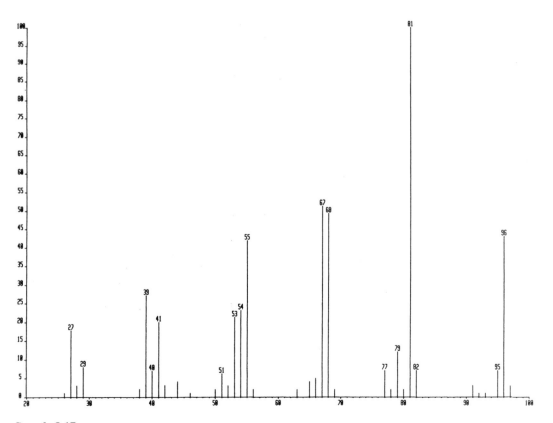

Sample 8.17

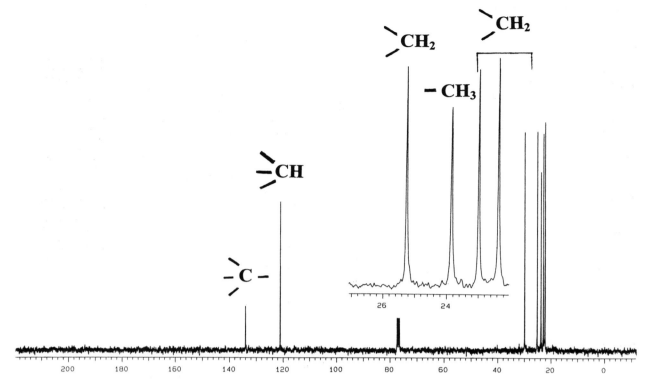

Sample 8.17 *Continued*

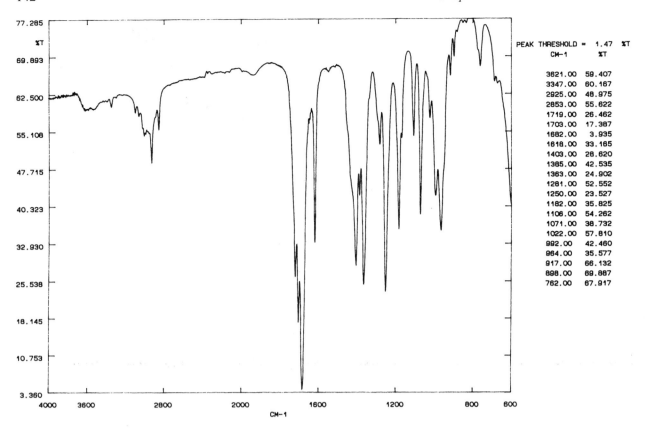

PEAK THRESHOLD =   1.47 %T

CM-1	%T
3621.00	59.407
3347.00	60.167
2925.00	48.975
2853.00	55.622
1719.00	26.462
1703.00	17.387
1682.00	3.935
1618.00	33.165
1403.00	28.620
1385.00	42.535
1363.00	24.902
1281.00	52.552
1250.00	23.527
1182.00	35.825
1106.00	54.262
1071.00	38.732
1022.00	57.810
992.00	42.460
964.00	35.577
917.00	66.132
898.00	69.887
762.00	67.917

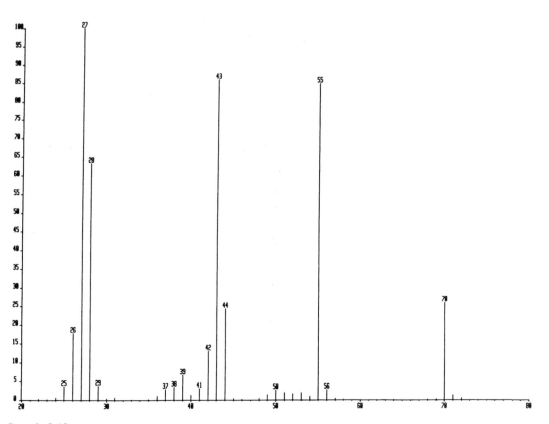

Sample 8.18

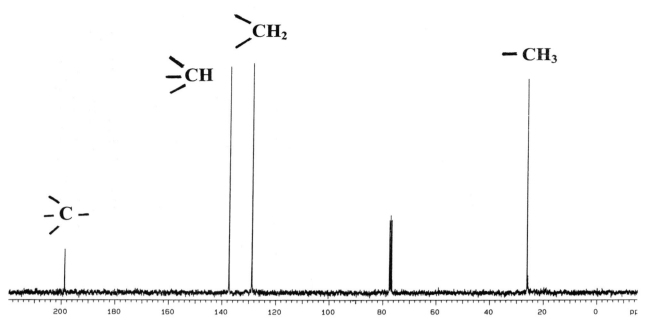

Sample 8.18 *Continued*

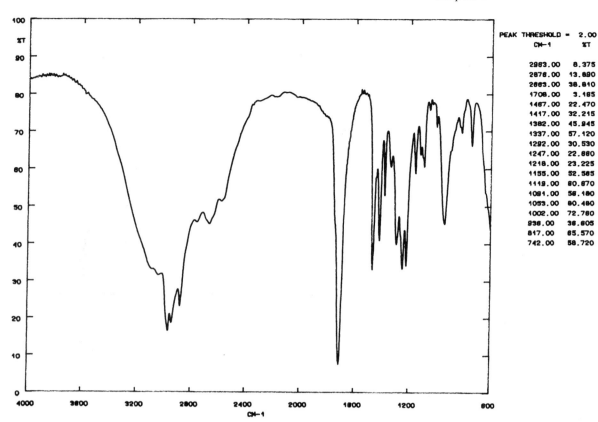

PEAK THRESHOLD = 2.00 %
CM-1   %T

2963.00   8.375
2876.00   13.890
2663.00   38.910
1708.00   3.165
1467.00   22.470
1417.00   32.215
1382.00   45.945
1337.00   57.120
1292.00   30.530
1247.00   22.660
1218.00   23.225
1155.00   52.585
1119.00   80.870
1091.00   56.180
1053.00   80.480
1002.00   72.780
936.00   38.605
817.00   65.570
742.00   58.720

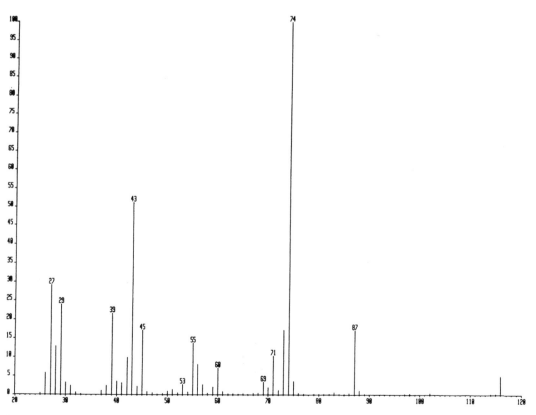

Sample 8.19

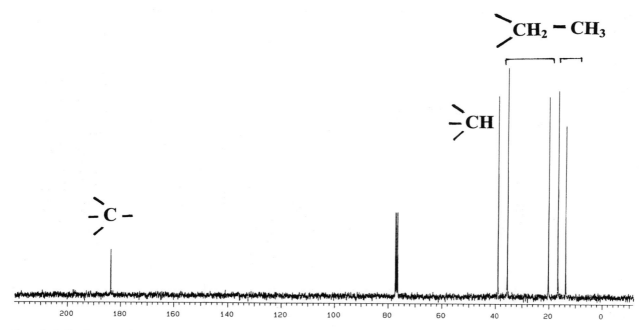

Sample 8.19 *Continued*

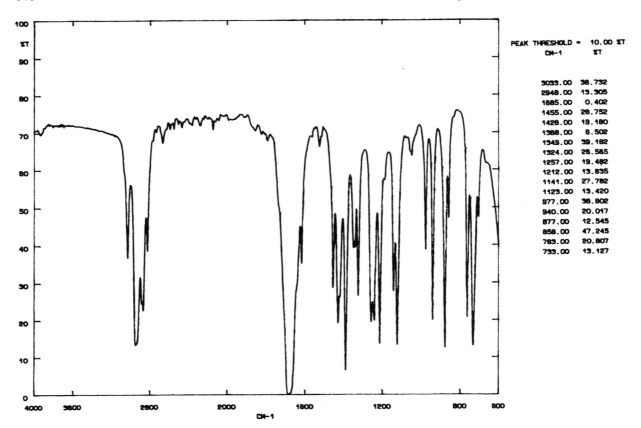

PEAK THRESHOLD = 10.00 %T

CM-1	%T
3033.00	36.732
2948.00	13.305
1685.00	0.402
1455.00	26.752
1428.00	19.180
1388.00	6.502
1349.00	39.182
1324.00	26.585
1257.00	19.482
1212.00	13.835
1141.00	27.782
1123.00	13.420
977.00	36.802
940.00	20.017
877.00	12.545
858.00	47.245
783.00	20.807
733.00	13.127

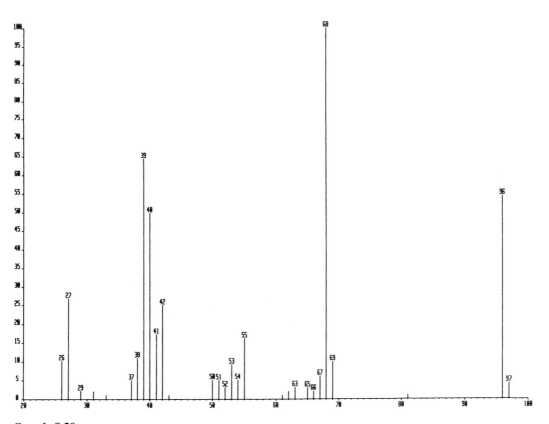

Sample 8.20

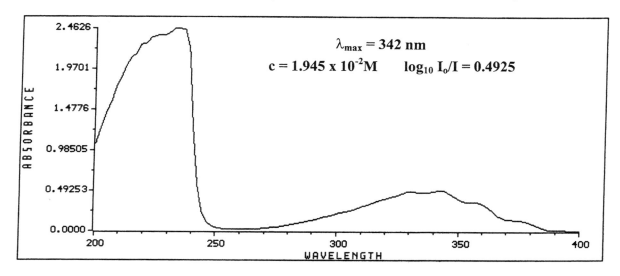

$\lambda_{max} = 342$ nm

$c = 1.945 \times 10^{-2}$M $\log_{10} I_0/I = 0.4925$

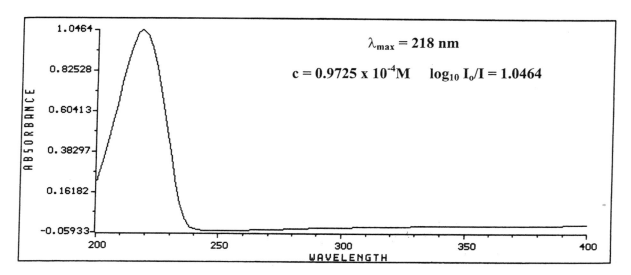

$\lambda_{max} = 218$ nm

$c = 0.9725 \times 10^{-4}$M $\log_{10} I_0/I = 1.0464$

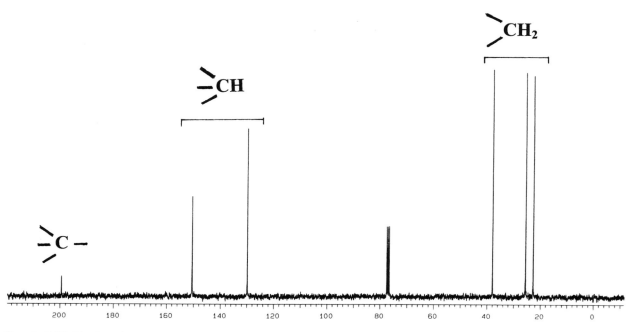

Sample 8.20 *Continued*

CHAPTER 9

^{1}H Nuclear Magnetic Resonance Spectroscopy

^{1}H nuclear magnetic resonance spectroscopy tells us about the environment of the hydrogen atoms in a molecule. The technique is based on exactly the same principles as ^{13}C NMR spectroscopy: the ^{1}H nucleus has nuclear spin $\frac{1}{2}$, so when placed in a strong magnetic field it can exist in higher or lower energy states. When the nucleus is irradiated, it absorbs radiofrequency radiation, and nucleii in lower energy spin states are promoted to higher energy spin states.

There are important differences between ^{1}H and ^{13}C NMR spectroscopy. The main one is that the ^{1}H atom, which has nuclear spin, comprises 99.98% of naturally occurring hydrogen. Consequently, most ^{1}H NMR spectra can be measured by a single scan, and Fourier transform methods are used only in exceptional circumstances. This means that peak areas are proportional to the number of hydrogen atoms that the peak represents, and this is very valuable when analysing spectra.

Another important difference is that since ^{12}C, the main carbon isotope, does not possess nuclear spin, most hydrogen atom spectra do not have any splitting from the carbon atom to which the hydrogen is attached. The few hydrogen atoms which are attached to ^{13}C atoms and hence are split produce tiny peaks which are usually lost in the background electronic noise on the spectrum. Consequently, we need consider only ^{1}H to ^{1}H coupling, unless we have present other atoms with nuclear spin such as fluorine or phosphorus.

As in ^{13}C NMR, the frequency of the radiation absorbed by a ^{1}H nucleus depends only on the magnetic field in which the ^{1}H nucleus is placed. When a spectrum is measured, this field has three components:

A. The main component, the applied field, which is very large, acts equally on all atoms.
B. The field resulting from the electron density around each nucleus. Movement of these electrons in the applied magnetic field generates a field which opposes the applied field, moving the absorption frequency of a neighbouring nucleus downfield. The effects produced on the ^{1}H nucleus absorption are similar to but smaller than those produced on ^{13}C absorption. The ^{13}C NMR absorption of neutral atoms covers a range of $\delta200$; the range of ^{1}H absorptions is approximately $\delta10$. Thus, the shift of absorption frequency, known as chemical shift, on a ^{1}H atom due to an electronegative neighbour is approximately 5% of the chemical shift on a similarly placed ^{13}C atom. A list of chemical shifts is given in Tables 9.1 and 9.2.

There is a problem with chemical shifts of the hydrogen atoms of groups such as –OH, –NH$_2$, –SH and –COOH, where the hydrogen atoms undergo rapid exchange with each other and with traces of water

149

Table 9.1 *^1H NMR chemical shifts of hydrogen atoms attached to carbon atoms in different environments*

| Group | ^1H chemical shift (δ) | |
	Attached to groups of low electronegativity	Full range
CH_3	0.9	0.9–3.8
CH_2	1.4	1.4–4.3
CH	1.5	1.5–4.5
C≡C–H	1.8	1.8–3.1
C=C–H	5.2	4.5–6.0
Ph	7.3	6.0–9.0
H–C=O	9.8	9.5–10

Table 9.2 *Effect of electronegative groups on ^1H NMR chemical shifts*

| Group | Chemical shift (δ) when attached to | | | | | |
	Alkyl	Aryl	C=O	C=C	–OH	–O–C=O
CH_3	0.9	2.3	2.0–2.2	2.0	3.5	3.7
CH_2	1.4	2.7	2.2–2.4	2.4	3.6	4.1
CH	1.5	3.0	2.5–2.7	2.7	3.9	4.8

present in the sample. The exchange is faster than the measurement of absorption of radiation, so we get a single peak for the exchanging atom and the water, whose chemical shift is a weighted average of the chemical shifts of the atoms involved. Such peaks can readily be identified by a procedure known as a 'D$_2$O shake'. A drop of deuterium oxide is added to the sample in the NMR tube, and the tube shaken briefly. The exchangeable hydrogen of the sample exchanges rapidly with deuterium. The H$_2$O generated, and the D$_2$O, float on top of the solvent, outside the region where the spectrum is being measured. The O<u>H</u> or similar peak thus disappears, as can be seen by comparing Figures 9.1 and 9.2.

The chemical shifts of O<u>H</u> groups do give some useful structure information. As a rough rule, alcohol OH peaks are found in the range $\delta 0$ to 5, phenolic OH peaks in the range $\delta 5$ to 10, and carboxylic acid COO<u>H</u> peaks in the range $\delta 10$ to 15.

C. The field resulting from the magnetic field of neighbouring nuclei. This has the effect of the signal from a ^1H atom being split by the magnetic fields of neighbouring nuclei, and provides the most valuable information from a ^1H NMR spectrum, as well as providing most of the problems.

The main ^1H to ^1H splitting of the absorption peak of a ^1H nucleus is by the magnetic fields of ^1H atoms on neighbouring carbon atoms. There can be small splittings from hydrogen atoms on the next but one carbon atom, but these are usually observed only in rigid systems, such as alkenes and cyclic systems. Splitting is always mutual; if H$_A$ splits H$_B$, then H$_B$ *must* split H$_A$.

Provided we have free rotation around the C–C bonds, the splitting patterns in ^1H to ^1H coupling are similar to those we have met when considering ^1H to ^{13}C coupling. The $(n+1)$ rule applies: a peak is split into $(n+1)$ peaks if it has n hydrogen atoms on neighbouring carbon atoms. Thus, a single hydrogen

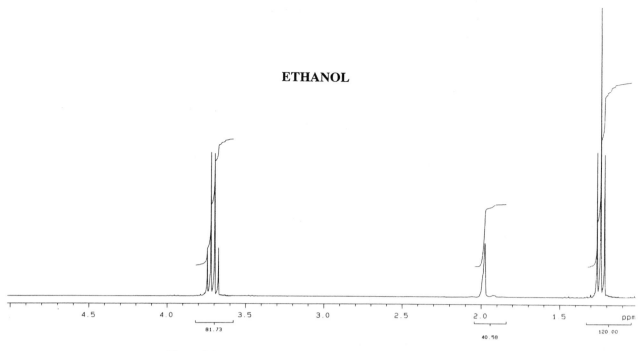

Figure 9.1

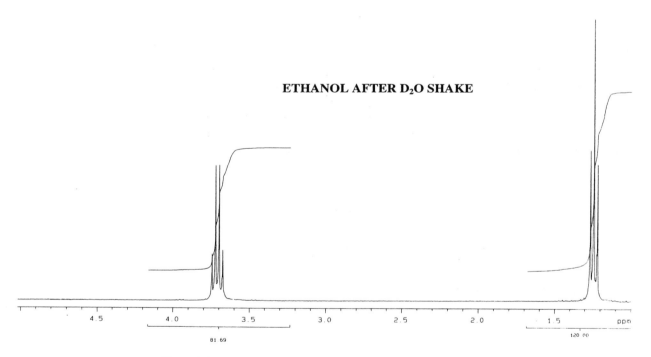

Figure 9.2

atom on a neighbouring carbon atom splits a ¹H signal into a doublet, two split it into a triplet, three into a quartet, up to the rare situation of nine atoms splitting a neighbouring atom absorption into ten peaks. Simple splittings are illustrated in Diagram 9.1.

These patterns depend on the assumption that the size of the splitting, expressed as the coupling constant, J, is the same for all couplings. The coupling constant depends on the dihedral angle between the carbon to

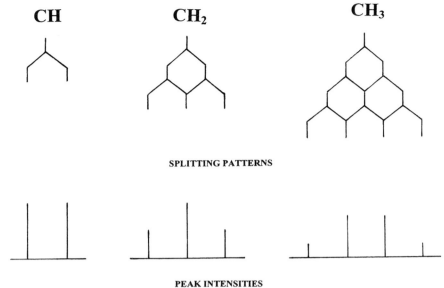

CH **CH₂** **CH₃**

SPLITTING PATTERNS

PEAK INTENSITIES

Diagram 9.1

hydrogen bonds of the two hydrogen atoms involved. Looking along the C–C bond, we can see this angle:

In any system in which we have free rotation of C–C bonds, the dihedral angle varies rapidly during measurement of absorption, so we see only an averaged value of J for all couplings and the patterns shown in Diagram 9.1 are observed. In rigid systems, J can vary over the range from 0 to 20 Hz and more complex patterns are observed. J values can be very useful, for example, in telling us if vicinal hydrogen atoms on a double bond are *cis* or *trans* to each other:

trans double bond
J_{AB} = 12–18 Hz

cis double bond
J_{AB} = 7–11 Hz

We also get geminal coupling if we have a =CH₂ group:

The J value is now dependent on the electronegativities of R^1 and R^2.

In rigid systems such as these, we can also get long range coupling:

J_{AB} = 3.5 Hz
J_{CB} = 3.0 Hz

J_{AB} = 1.8 Hz

The coupling constant is easily measured: it is the difference, in Hz, of the frequencies of absorption of the peaks produced by the splitting. It is independent of the applied field. If we have a regular multiplet, as shown in Diagram 9.1, then all the coupling constants are the same, and are given by the difference in absorption frequency of any pair of neighbouring peaks in the multiplet.

Like ^{13}C NMR spectra, ^1H NMR spectra are measured in solution, usually in $CDCl_3$, since it does not contain any hydrogen and so does not produce a ^1H NMR signal. Any residual $CHCl_3$ produces a peak at $\delta7.25$. The solution is placed in a thin walled glass tube, and spun in a powerful magnetic field. The absorption of radiofrequency radiation is then measured. A typical spectrum is that of ethanol, CH_3CH_2OH, shown in Figure 9.1. The positions of the peaks can be read from the scale; always read to the centre of a multiplet. Many spectrometers print the peak position in Hz or δ, above the peak, which is very valuable when measuring coupling constants, but chemical shifts can be read from the scale. Note that the peaks are cut by flattened S-shaped curves; the height between the horizontal parts of the curve is proportional to the area enclosed by the peaks, which is itself proportional to the number of hydrogen atoms producing the peak. The area is also measured electronically, and recorded below the δ scale. Measurements of peak area are not precise, since they also include electronic noise and impurity peaks. They give the ratio of peak areas, which must be related to the hydrogen content of the sample. In this case, the ratio is 82.48 to 40.14 to 120.00, which in a molecule with six hydrogen atoms, makes the signals represent 2, 1 and 3 hydrogen atoms respectively.

The spectrum shown in Figure 9.1 consists of a quartet at $\delta3.71$ representing two hydrogen atoms, a single peak at $\delta1.98$ representing one hydrogen atom, and a triplet at $\delta1.24$ representing three hydrogen atoms. Figure 9.2 shows that the peak at $\delta1.98$ disappears after a D_2O shake, so it is clearly the O$\underline{H}$ hydrogen. The C$\underline{H}_2$ peak at $\delta3.71$ is a quartet, so must be next to a CH_3 group, while the $-C\underline{H}_3$ peak at $\delta1.24$ is a triplet, so must be next to a CH_2 group. Clearly, the spectrum of ethanol can be assigned:

$$CH_3–CH_2–OH$$
$$\delta \quad 1.24 \quad 3.71 \quad 1.98$$

It is very important to note that the CH_3 group which absorbs at $\delta1.24$ is a triplet because it is next to a CH_2, and NOT because it has three hydrogen atoms on one carbon atom. The signal splitting depends on hydrogen atoms on NEIGHBOURING carbon atoms.

The next spectrum to consider is that of 1-methylethanol [propan-2-ol, $(CH_3)_2CHOH$], shown in Figure 9.3. The spectrum consists of a multiplet at $\delta4.02$, a broad singlet at $\delta1.88$, and a doublet at $\delta1.20$. The peak at $\delta1.88$ disappears on a D_2O shake, so is clearly the O$\underline{H}$ peak. The doublet at $\delta1.20$ represents six hydrogen atoms, whilst the multiplet at $\delta4.02$ represents one hydrogen atom and is moved downfield by being next to an electronegative group, probably the hydroxyl. The doublet is from six hydrogen atoms on carbon atoms next to a carbon atom with one hydrogen atom. The multiplet at $\delta4.02$ looks like a quintet, but since it is on a carbon atom next to carbons carrying six hydrogen atoms, it must be a septet. Deciding the exact multiplicity of multiplets of low intensity is often difficult, but an amplified expansion can help. The spectrum is clearly that of a molecule with a C$\underline{H}$ group attached to two C$\underline{H}_3$ groups and an O$\underline{H}$ group:

$$(CH_3)_2–CH–OH$$
$$\delta \quad 1.20 \quad 4.02 \quad 1.88$$

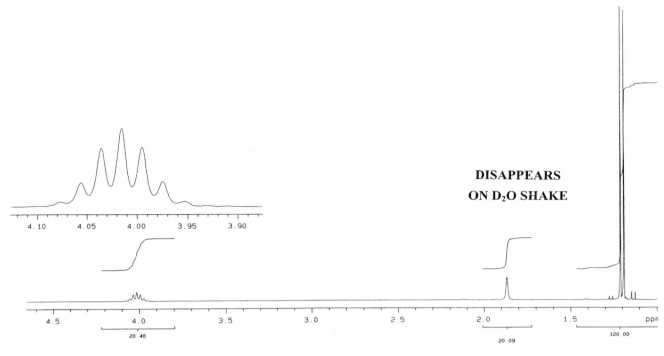

Figure 9.3

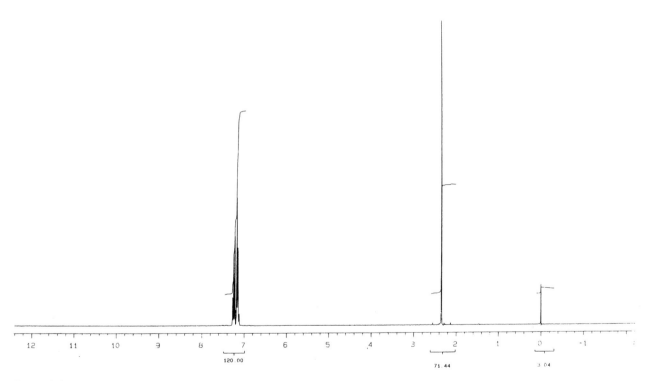

Figure 9.4

The next spectrum, shown in Figure 9.4, is that of toluene, $PhCH_3$. The singlet at $\delta 2.35$ results from the methyl group, and is moved downfield from the previous methyl groups by the effect of the phenyl ring. The aromatic hydrogen atoms give rise to a complex multiplet around $\delta 7.2$. The chemical shifts of the aromatic hydrogen atoms depend on their distance from the CH_3 group, and they split each other to give a complex system which we need not assign.

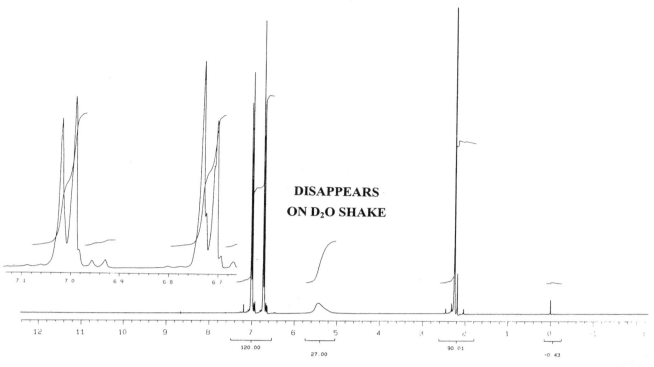

DISAPPEARS
ON D₂O SHAKE

Figure 9.5

Despite the complexity of the aromatic region, we can often get useful information about the arrangement of substituents around the ring in disubstituted aromatic compounds by considering the symmetry, or lack of it, in the ring. Figure 9.5 shows the spectrum of a 1,4- or *para*-substituted compound, 4-methylphenol. The pattern of four peaks between $\delta6.7$ and 7.1 is typical of a *para*-substituted compound. The molecule has a plane of symmetry, so that the hydrogen atom on C-2 is coupled to the hydrogen atom on C-3, giving a pair of doublets identical to those produced by the hydrogen atoms on C-6 and C-5. The peak at $\delta5.4$ results from the phenolic O<u>H</u>, while the peak at $\delta2.25$ results from the methyl group. Small peaks near the main peak denote the presence of impurities. Figure 9.6 shows the spectrum of the 1,3- or *meta*-substituted isomer, 3-methylphenol. In a *meta*-substituted compound, the single hydrogen atom on the carbon atom between the substituents gives an unspilt peak, except for the small splittings arising from long range coupling. In this case, the peak from the C-2 hydrogen is at $\delta6.64$. The spectrum of the 1,2- or *ortho*-substituted isomer, 2-methylphenol, is shown in Figure 9.7. In this case, two hydrogen atoms give doublets and two hydrogen atoms give triplets. The chemical shifts are often not enough to separate the peaks, and the result is frequently complex. If the substituents are identical, the spectra of all the disubstituted aromatics are greatly simplified.

Before we start to consider the interpretation of ^1H NMR spectra, two other spectra need to be considered, as they illustrate problems which are not readily predictable. The first of these is shown in Figure 9.8 and is the spectrum of 1,2-dimethoxyethane, $CH_3-O-CH_2-CH_2-O-CH_3$. It consists of two singlets, at $\delta3.37$ and $\delta3.51$. The methyl peaks, at $\delta3.37$, would not be expected to shown any splitting, but the CH_2 peaks, at $\delta3.51$, are next to each other. However, it is always found that hydrogen atoms which are in identical environments do not split each other. We have no coupling, and hence no splitting of the CH_2 groups which are in identical environments. The second spectrum is that shown in Figure 9.9, which is the spectrum of *N,N*-

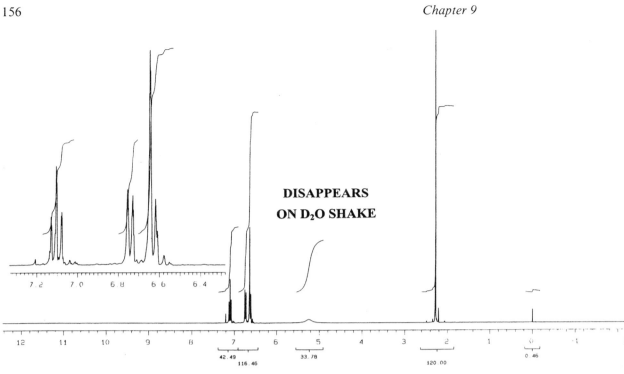

Figure 9.6

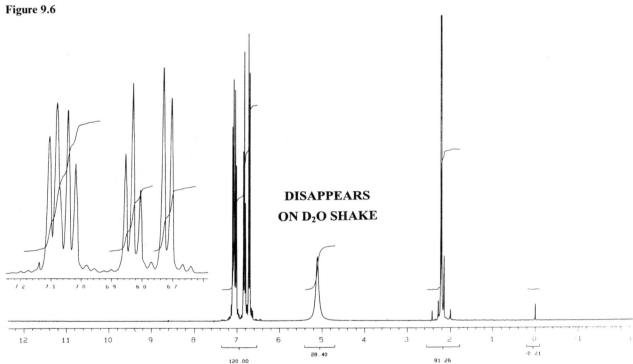

Figure 9.7

dimethylformamide, $(CH_3)_2NCHO$. The spectrum shows two signals at $\delta 2.88$ and $\delta 2.96$, which at first glance might be through to be a CH_3 group next to a CH. However, we do not have any multiplets in the spectrum, so that separation of the signals cannot be the result of splitting, and must result from chemical shift. Although both methyl groups are attached to the same nitrogen atom, there is restricted rotation around the N–C bond, and the two methyl groups are in different positions relative to the carbonyl group, and hence have different chemical shifts.

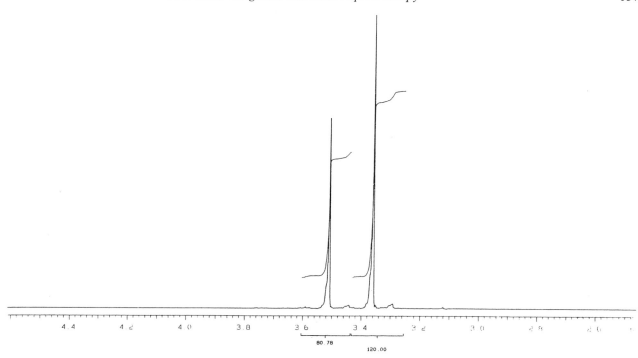

Figure 9.8

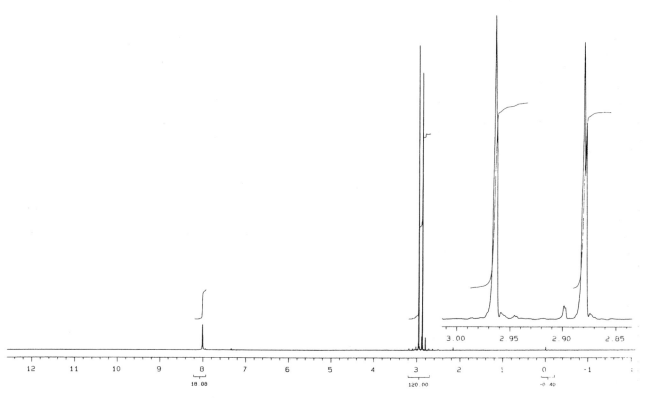

Figure 9.9

We can now look at unknown spectra. The first of these, shown in Figure 9.10, is of a substance $C_4H_8O_2$. When we look at the spectrum, we can see that it consists of a quartet at $\delta 4.12$, a singlet at $\delta 2.05$ and a triplet at $\delta 1.27$. The peaks all show very small extra splittings, due to long range coupling, which in this case are unimportant. The three groups of peaks have peak areas in the

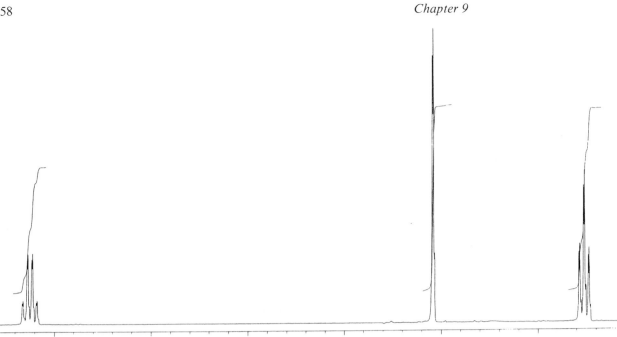

Figure 9.10

ratio 81.46 to 120.00 to 117.06, which is approximately 2 : 3 : 3. Since we have eight hydrogen atoms in the molecule, we must have a CH_2 group absorbing at $\delta 4.12$, a CH_3 group at $\delta 2.05$ and another CH_3 group at $\delta 1.27$. The chemical shifts show that the hydrogen atoms giving the peak at $\delta 4.12$ are probably next to a strongly electronegative group, and the hydrogen atoms giving the peak at $\delta 2.05$ are next to a weaker electronegative group or further away from a strongly electronegative group. The CH_2 group giving rise to the quartet at $\delta 4.12$ must be next to a CH_3 group, and the CH_3 group giving rise to the triplet at $\delta 1.27$ must be next to a CH_2 group. Clearly, these form an ethyl group, attached to an electronegative atom or group. The CH_3 group giving rise to the peak at $\delta 2.05$ is not next to a hydrogen bearing carbon atom. We must now look at the original molecular formula, $C_4H_8O_2$. We have accounted for C_2H_5 plus CH_3, so we have COO unaccounted for. This is most likely to be an ester group. Since the CH_2 of the ethyl group is moved down to $\delta 4.12$, it must be attached to the oxygen atom. The unsplit methyl group at $\delta 2.05$ must be attached to the carbonyl group, which moves it a shorter distance downfield. The substance is thus ethyl acetate:

$$CH_3-CO-O-CH_2-CH_3$$
$$\delta \quad 2.05 \qquad\qquad 4.12 \quad 1.27$$

The next substance, whose ^1H NMR spectrum is shown in Figure 9.11, has the molecular formula C_4H_9Br. The spectrum shows four groups of peaks, plus added TMS at $\delta 0.0$. The groups of peaks are at $\delta 3.41$, 1.85, 1.47 and 0.94, in the ratio 83.15 to 81.47 to 84.01 to 120.00, which is approximately 2:2:2:3. Combining this data with the molecular formula, we have three CH_2 groups and a CH_3 group. The CH_2 group absorbing at $\delta 3.41$ is probably downfield because it is attached to the bromine atom. Thus, we have now accounted for all the atoms listed in the molecular formula. Next, we must consider the splitting patterns. The peaks at $\delta 3.41$ form a triplet, those at $\delta 1.85$ a quintet.

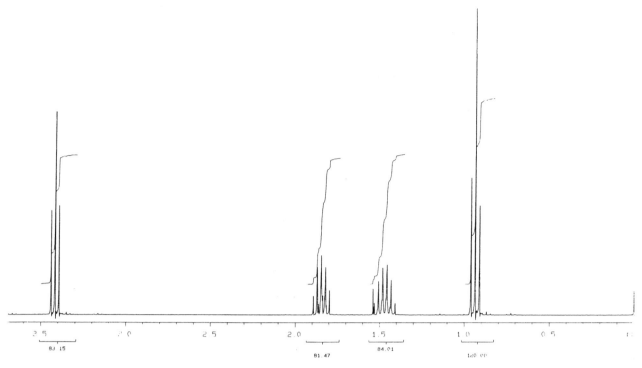

Figure 9.11

At $\delta 1.47$ we have a symmetrical sextet, with a small extra peak at $\delta 1.54$. Since this does not fit with an otherwise symmetrical group, and is far too small to represent even a single hydrogen, it is probably an impurity. At $\delta 0.94$, we have a triplet. It is often convenient to summarise the data:

Peak at δ	No. of H	Next to
3.41	2	2H
1.85	2	4H
1.47	2	5H
0.94	3	2H

We next seek to assemble the groups into a molecule. We can do this by using the diminishing effect on chemical shift of the bromine atom, or we can use the splitting patterns. We have two 'end of chain' groups, CH_2Br and CH_3, so we have a straight chain molecule. Assembly can start at either end of the chain. Starting from CH_2Br, at $\delta 3.41$, we see that it is a triplet, so must be next to a CH_2. We have two further CH_2 groups, one a quintet, the other a sextet. If we put the latter next to our CH_2Br, then it would also have to have a CH_3 group attached, giving $CH_3CH_2CH_2Br$, which would leave us with an unused CH_2 group. However, if we join the CH_2 group which absorbs at $\delta 1.85$, to our CH_2Br, then it would have to be attached to another CH_2, since it is next to carbon atoms carrying four hydrogens. This has to be the only remaining CH_2, which absorbs at $\delta 1.47$, and since it has to be next to carbon atoms carrying a total of five hydrogens, must be next to the only group left, the CH_3.

If we started at the other end of the chain, with the CH_3 group, then this must be next to a CH_2. If it were attached to the CH_2 which gives a quintet at $\delta 1.85$, then the next group along the chain would have to be a CH group,

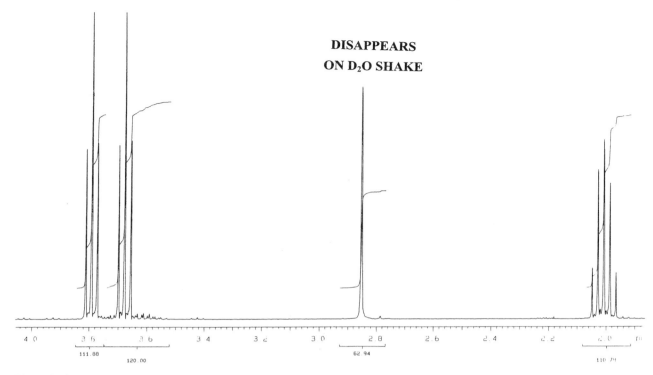

Figure 9.12

which we do not have. It must therefore be attached to the CH_2 group which gives a sextet at $\delta 1.47$, which must be next to another CH_2, clearly that absorbing at $\delta 1.85$. This needs to be attached to another CH_2, clearly the chain ending group at $\delta 3.41$. Either approach gives us:

$$Br{-}CH_2{-}CH_2{-}CH_2{-}CH_3$$

δ	3.41	1.85	1.47	0.94
next to	2H	4H	5H	2H

Our assignment fits with the expected decrease in chemical shift as we move away from the electronegative bromine.

The next spectrum, shown in Figure 9.12, is of a substance C_3H_7OCl. The spectrum consists of peaks at $\delta 3.79$ (2H, triplet), 3.68 (2H, triplet), 2.85 (1H, singlet) and 2.01 (2H, quintet). The peak at $\delta 2.85$ disappears on shaking the sample with D_2O, so the hydrogen is probably attached to the oxygen atom. Thus we have

Peak at δ	No. of H	Next to
3.79	2	2H
3.68	2	2H
2.01	2	4H

The peaks at $\delta 3.79$ and 3.68 are moved downfield, by the electronegative OH and Cl respectively. The molecule thus consists of three CH_2 groups, so that the middle CH_2 is a quintet, while the outer CH_2 groups have a chlorine attached to one, and a hydroxyl attached to the other:

$$Cl{-}CH_2{-}CH_2{-}CH_2{-}OH$$

δ	3.68	2.01	3.79	2.85
next to	2H	4H	2H	

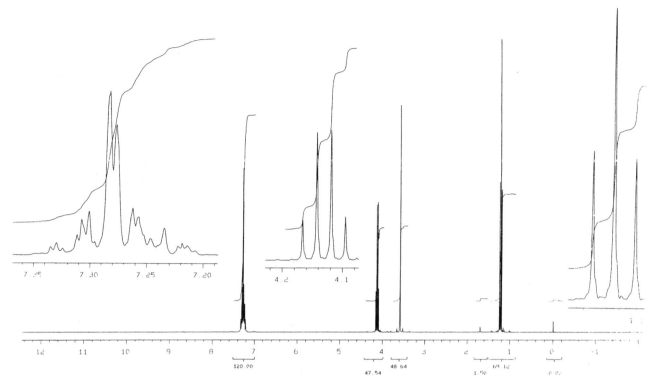

Figure 9.13

The next spectrum, shown in Figure 9.13, is of a substance $C_{10}H_{12}O_2$. Parts of the spectrum have been enlarged to help interpretation. The spectrum shows peaks at $\delta 7.30$ (5H, complex multiplet), 4.15 (2H, quartet), 3.58 (2H, singlet) and 1.22 (3H, triplet). Thus, we have

Peak at δ	No. of H	Next to
7.30	5	–
4.15	2	3H
3.58	2	Zero H
1.22	3	2H

The multiplet at $\delta 7.30$ is clearly from an aromatic ring. It represents 5H, so the ring must be monosubstituted. The peaks at $\delta 4.15$ and 1.22 represent an ethyl group next to an electronegative atom, and the peak at $\delta 3.58$ a CH_2, which has no hydrogen atoms on neighbouring carbon atoms, is next to one or two electronegative groups. These groups, C_6H_5 plus C_2H_5 plus CH_2, account for C_9H_{12} of the $C_{10}H_{12}O_2$ of the original substance, so we have CO_2 left over; this is probably an ester linkage. The ethyl group must be attached to the oxygen atom of the ester, so that the CH_2 links the carbonyl group to the chain ending phenyl group:

$$\begin{array}{c} \quad\quad\quad\quad\quad O \\ \quad\quad\quad\quad\quad \| \\ Ph{-}CH_2{-}C{-}O{-}CH_2{-}CH_3 \end{array}$$

δ	7.30	3.58	4.15	1.22
next to			3H	2H

The final example has the spectrum shown in Figure 9.14, and has the molecular formula $C_5H_{10}O$. The spectrum shows peaks at $\delta 5.40$ (1H, broad

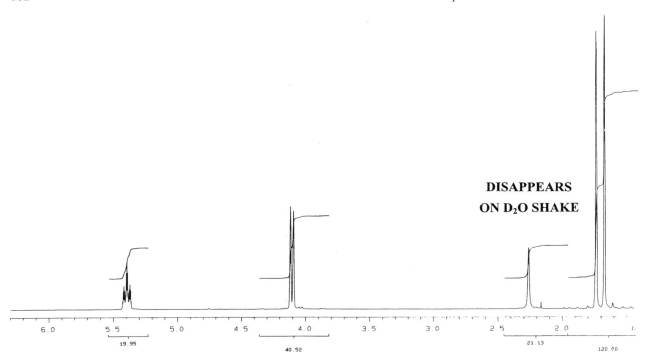

Figure 9.14

triplet), 4.10 (2H, doublet), 2.27 (1H, singlet), 1.75 (3H, singlet) and 1.68 (3H, singlet). We know that the peaks at $\delta 1.75$ and 1.68 are separate peaks, rather than a single split peak, because their separation comfortably exceeds any other splitting in the spectrum. It is an unbreakable rule of ^{1}H NMR spectroscopy that if H_A splits H_B, then H_B splits H_A, and the splittings are the same size. The peak at $\delta 5.40$ (1H) probably results from an H atom attached to a C=C double bond. The peak at $\delta 2.27$ disappears on a D_2O shake, so the hydrogen is attached to the oxygen in the molecule, and this OH is probably responsible for the downfield shift of the CH_2 group to $\delta 4.10$. We thus have:

Peak at δ	No. of H	Next to
5.40	1	2H
4.10	2	1H
1.75	3	–
1.68	3	–

The molecule must consist of a double bond to which H, CH_2OH, CH_3 and CH_3 are all attached. However, the splitting between the peaks at $\delta 5.40$ and 4.10 suggests they are on neighbouring carbon atoms, so the structure must be:

$$HO-CH_2-CH=C\begin{smallmatrix}CH_3\\CH_3\end{smallmatrix}$$

δ	2.27	4.10	5.40	1.75 and 1.68
next to		H	2H	

The extra line broadening on the peaks at $\delta 5.40$ probably results from a long range interaction with the methyl groups. Note that when we encountered the

^{13}C NMR spectrum of this molecule in Chapter 5, we were unable to prove the structure. Interaction between hydrogen atoms on neighbouring carbon atoms now shows us which carbon atom is joined to which.

Summary

To determine a structure from a ^{1}H NMR spectrum, we usually need the molecular formula. Analysis of the spectrum involves:

A. Look at the integration, and from the molecular formula see how many hydrogen atoms are represented by each group of signals.

B. Look for any signals which disappear on a D_2O shake.

C. See how many carbon atoms are involved in the groups of hydrogen atoms. Then look at the molecular formula to see how many carbon atoms are left over, and which other atoms the molecule contains.

D. Look at the chemical shifts. This should identify alkenes, alkynes and aromatic groups, and show which atoms are moved downfield by neighbouring electronegative substituents.

E. Look at the splittings of the peaks, and hence determine how many hydrogen atoms are on the carbon next to that carrying the hydrogen atom(s) that you are studying.

F. Assemble the atoms into a molecule. With short carbon chains, it will be obvious how the carbon atoms fit together, but for longer chains, work from a 'chain ending group' (*i.e.* a group with only one unused linkage), such as CH_3, CH_2OH, CH_2Br, $=CH_2$, phenyl, *etc*. If you have more than two of these groups, the molecule must be branched.

Problems in interpreting ^{1}H NMR spectra

You are given the ^{1}H NMR spectra and molecular formulae of 10 substances. Assign all the structures from the data.

$C_{10}H_{15}N$

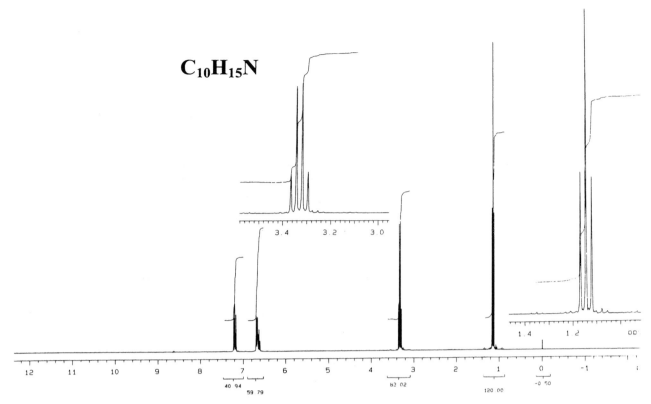

Sample 9.1

$C_5H_{10}O_2$

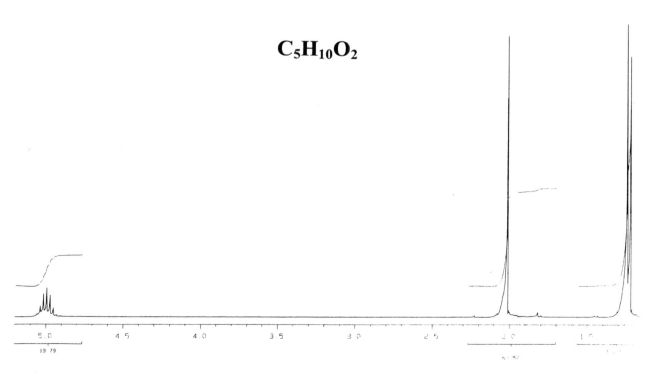

Sample 9.2

C₂H₅OBr

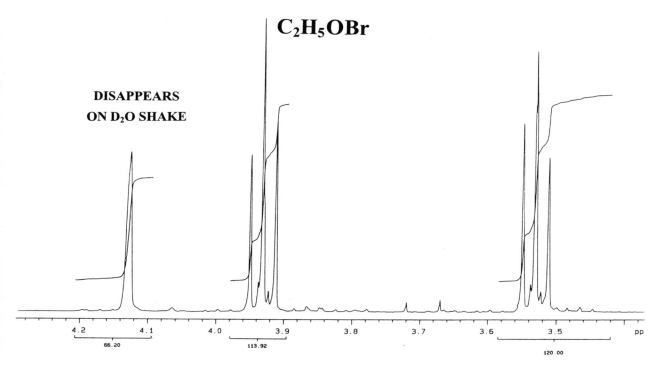

DISAPPEARS
ON D₂O SHAKE

Sample 9.3

C₈H₁₀O

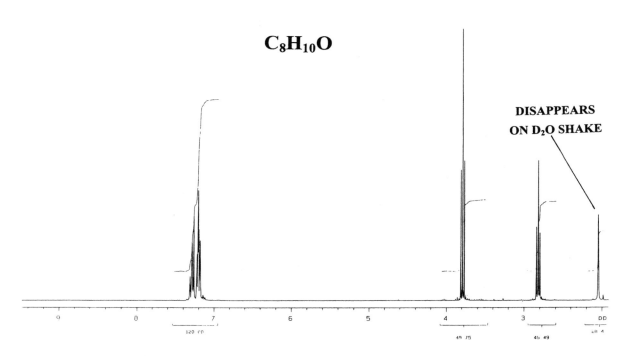

DISAPPEARS
ON D₂O SHAKE

Sample 9.4

C₃H₃O₂Cl

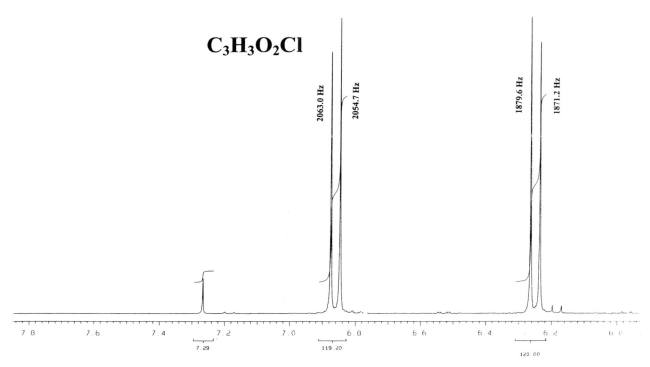

Sample 9.5

C₈H₈O₂

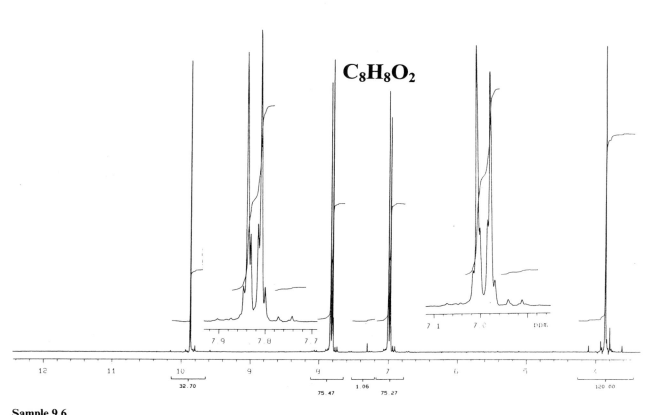

Sample 9.6

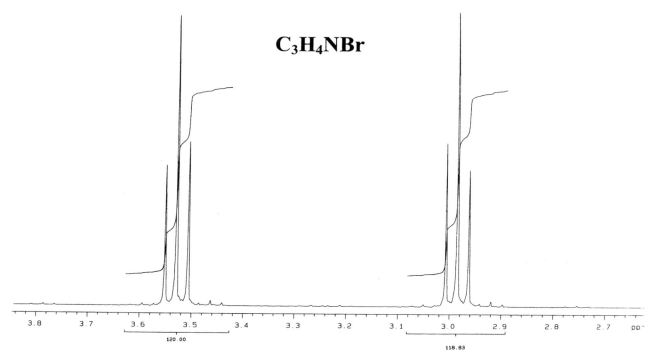

C₃H₄NBr

Sample 9.7

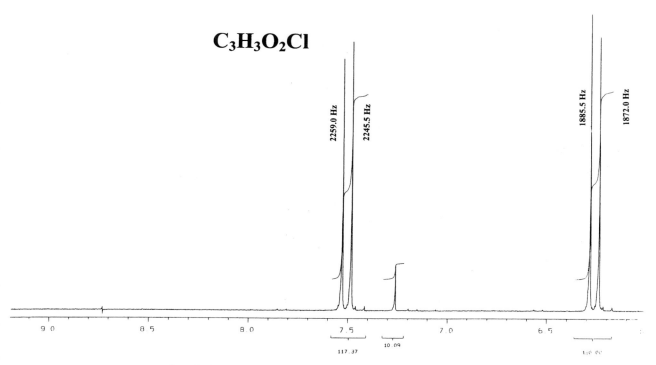

C₃H₃O₂Cl

Sample 9.8

C₉H₁₀O

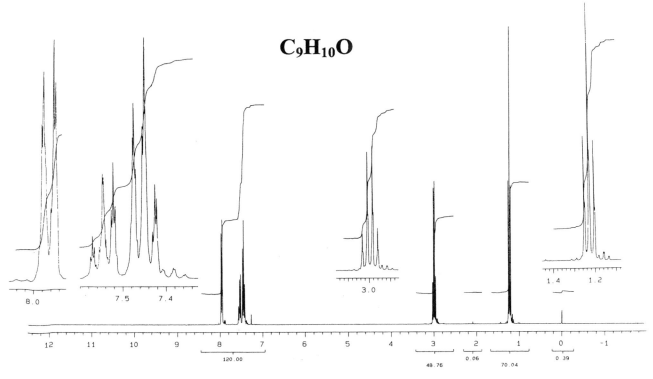

Sample 9.9

C₇H₅O₄N

**DISAPPEARS
ON D₂O SHAKE**

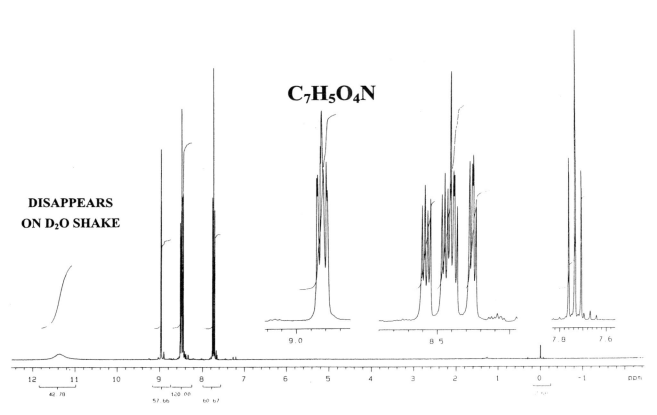

Sample 9.10

CHAPTER 10

Problems in Interpreting Infrared and ^{1}H Nuclear Magnetic Resonance Spectra

^{1}H NMR spectroscopy tells us very little about the functional groups present in a molecule. We can detect hydrogen atoms on carbonyl groups, on aromatic rings, on double bonds, on triple bonds and on atoms such as oxygen, nitrogen and sulfur. Even this is more than we were able to manage with ^{13}C NMR spectroscopy. Consequently, ^{1}H NMR and infrared spectroscopy provide a valuable partnership, one identifying the functional groups and the other the structure. Even so, we need to have the molecular formula provided, as neither technique will give conclusive evidence for the presence of halogens, and it is essential to know how many hydrogen atoms a molecule has before we attempt to assign the ^{1}H NMR spectrum.

In this exercise, all the necessary data are provided for 10 substances. You should be able to identify them from this information.

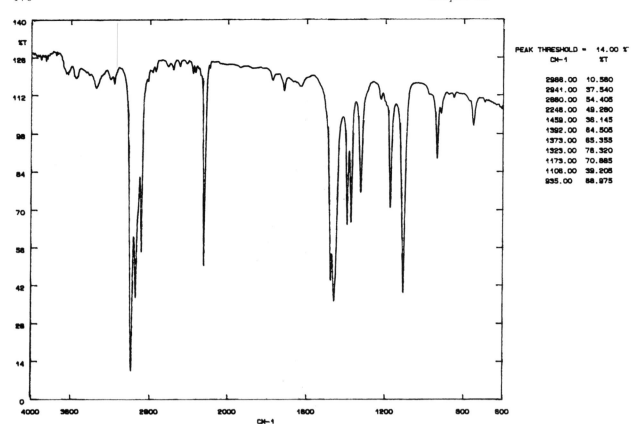

PEAK THRESHOLD = 14.00 %

CM-1	%T
2988.00	10.580
2941.00	37.540
2880.00	54.405
2248.00	49.280
1459.00	38.145
1392.00	84.505
1373.00	85.355
1323.00	78.320
1173.00	70.885
1108.00	39.205
935.00	88.975

C₄H₇N

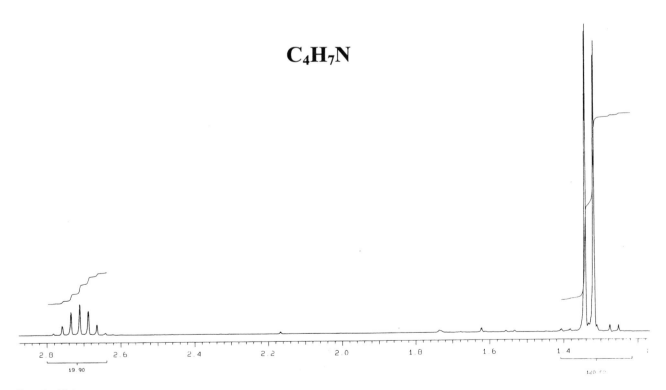

Sample 10.1

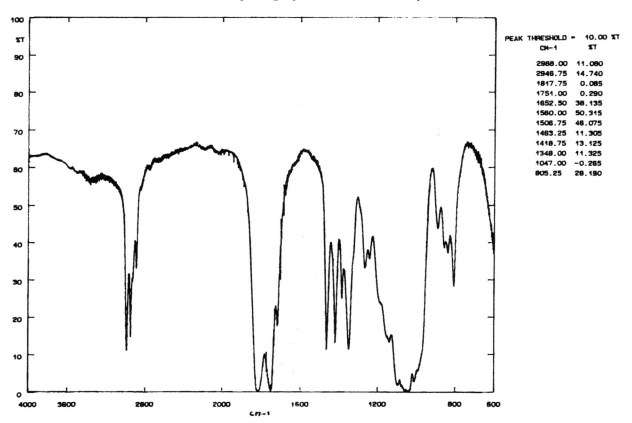

PEAK THRESHOLD =   10.00 %T

CM-1	%T
2988.00	11.080
2946.75	14.740
1817.75	0.085
1751.00	0.290
1652.50	38.135
1560.00	50.315
1508.75	46.075
1483.25	11.305
1418.75	13.125
1348.00	11.325
1047.00	-0.285
805.25	28.190

$C_8H_{14}O_3$

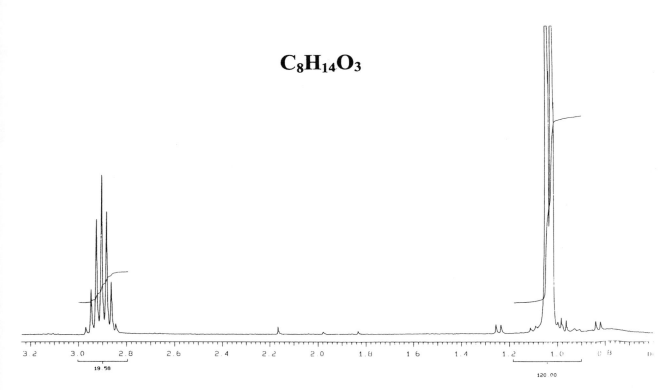

Sample 10.2

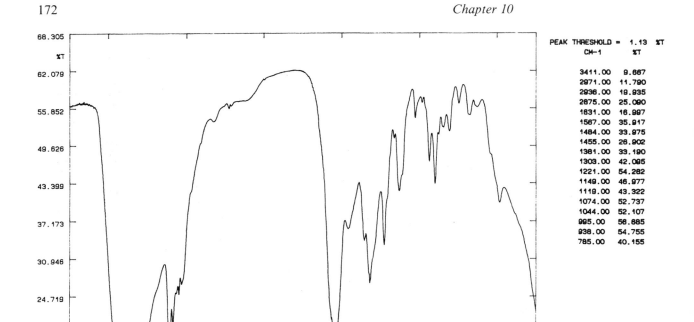

PEAK THRESHOLD = 1.13 %T

CM-1	%T
3411.00	9.667
2971.00	11.790
2936.00	19.935
2875.00	25.090
1831.00	16.997
1567.00	35.917
1484.00	33.975
1455.00	26.902
1381.00	33.190
1303.00	42.095
1221.00	54.282
1149.00	46.977
1119.00	43.322
1074.00	52.737
1044.00	52.107
995.00	56.685
938.00	54.755
785.00	40.155

$C_4H_{11}N$

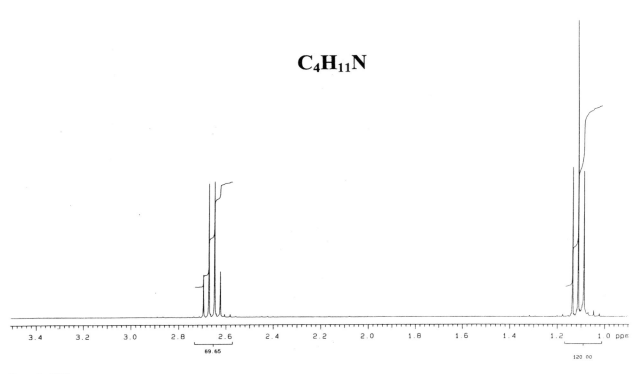

Sample 10.3

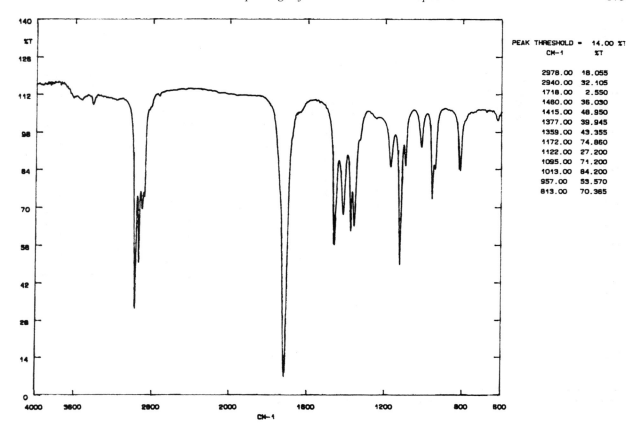

CM-1	%T
2978.00	18.055
2940.00	32.105
1718.00	2.550
1480.00	36.030
1415.00	48.950
1377.00	39.945
1359.00	43.355
1172.00	74.860
1122.00	27.200
1095.00	71.200
1013.00	84.200
957.00	53.570
813.00	70.365

$C_5H_{10}O$

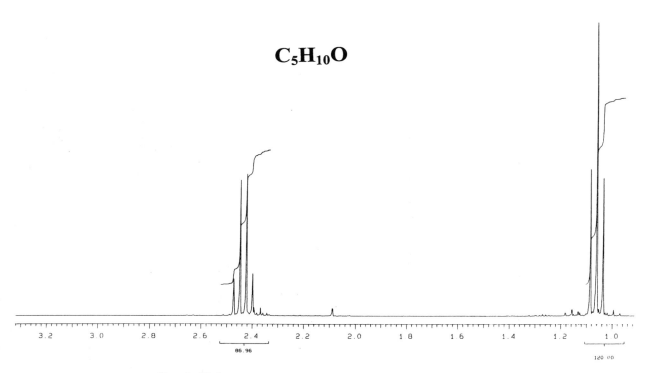

Sample 10.4

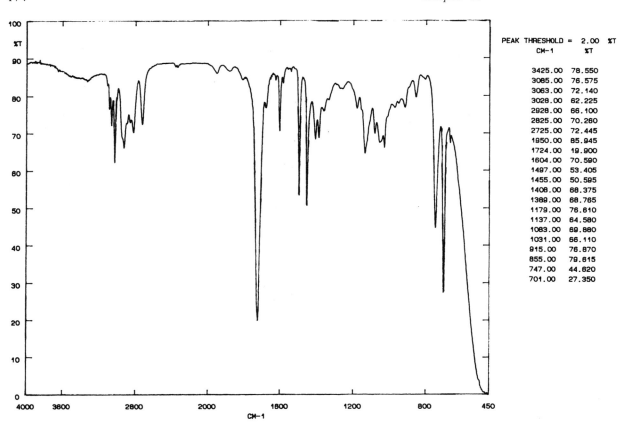

PEAK THRESHOLD =   2.00 %T
CM-1	%T
3425.00	78.550
3085.00	76.575
3063.00	72.140
3028.00	62.225
2928.00	66.100
2825.00	70.260
2725.00	72.445
1950.00	85.945
1724.00	19.900
1604.00	70.590
1497.00	53.405
1455.00	50.595
1408.00	68.375
1389.00	68.765
1179.00	76.610
1137.00	64.580
1083.00	69.880
1031.00	66.110
915.00	76.870
855.00	79.615
747.00	44.620
701.00	27.350

C₉H₁₀O

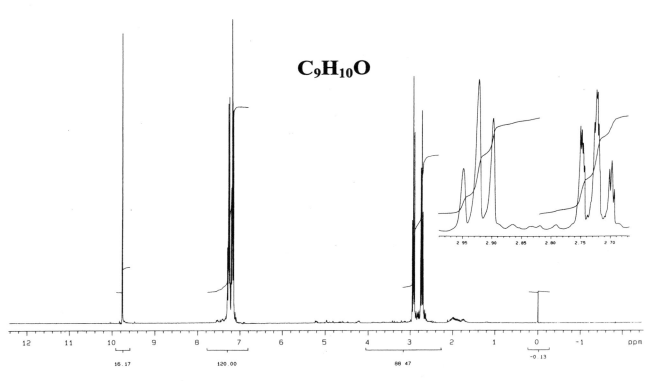

Sample 10.5

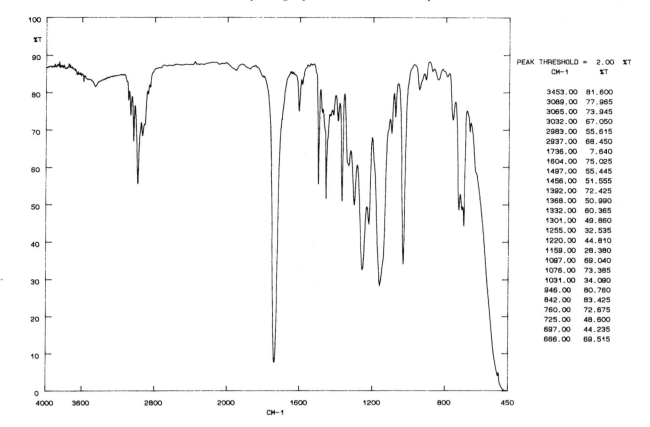

PEAK THRESHOLD = 2.00 %T

CM-1	%T
3453.00	81.600
3089.00	77.965
3065.00	73.945
3032.00	67.050
2983.00	55.615
2937.00	68.450
1736.00	7.640
1604.00	75.025
1497.00	55.445
1456.00	51.555
1392.00	72.425
1368.00	50.990
1332.00	60.365
1301.00	49.860
1255.00	32.535
1220.00	44.810
1159.00	28.380
1097.00	69.040
1076.00	73.385
1031.00	34.090
946.00	80.760
842.00	83.425
760.00	72.675
725.00	48.600
697.00	44.235
666.00	69.515

$C_{10}H_{12}O_2$

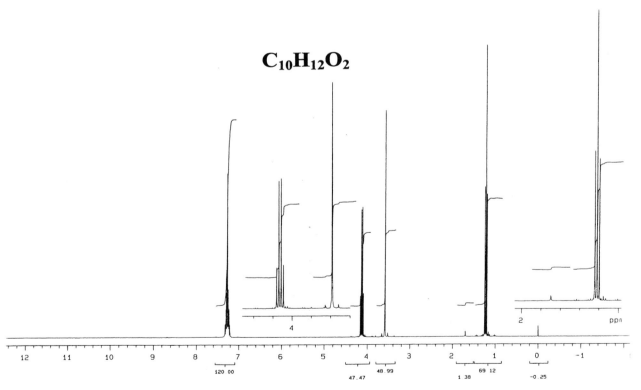

Sample 10.6

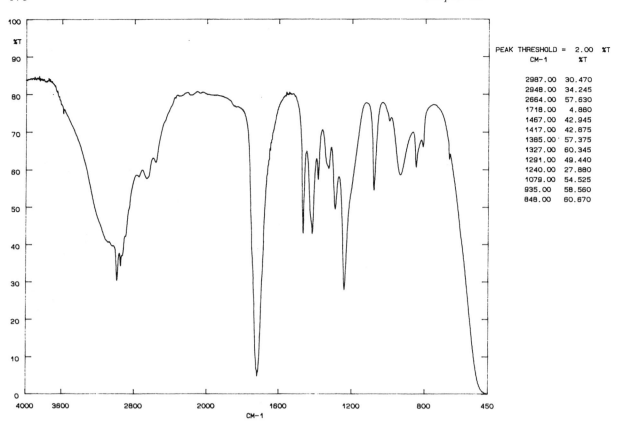

PEAK THRESHOLD = 2.00 %T
   CM-1   %T

   2987.00   30.470
   2948.00   34.245
   2664.00   57.630
   1718.00   4.880
   1467.00   42.945
   1417.00   42.875
   1385.00   57.375
   1327.00   60.345
   1291.00   49.440
   1240.00   27.880
   1079.00   54.525
    935.00   58.560
    848.00   60.670

C₃H₆O₂

**DISAPPEARS
ON D₂O SHAKE**

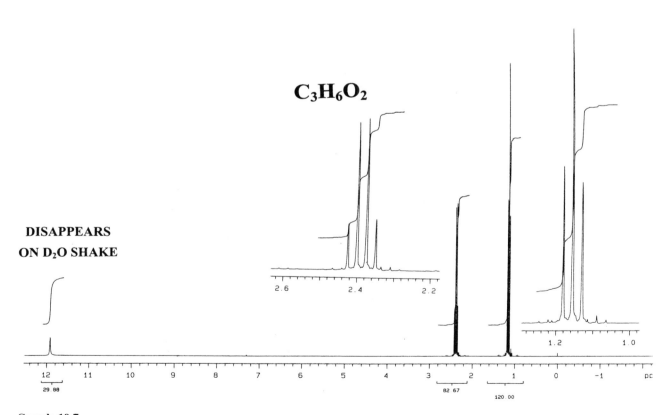

Sample 10.7

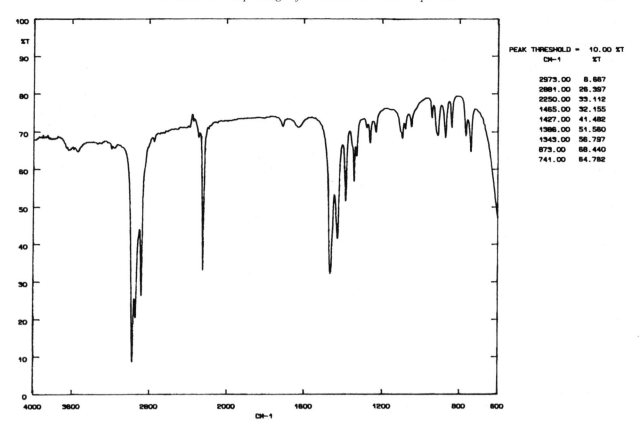

PEAK THRESHOLD =   10.00 %T
       CM-1   %T

      2973.00   8.687
      2881.00   28.397
      2250.00   33.112
      1465.00   32.155
      1427.00   41.482
      1386.00   51.560
      1343.00   56.797
       873.00   88.440
       741.00   84.782

$$C_4H_7N$$

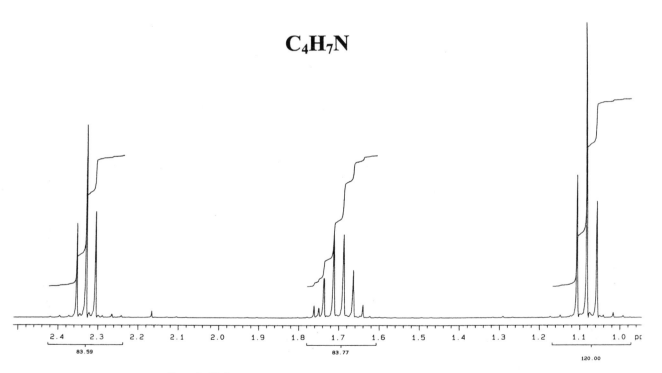

Sample 10.8

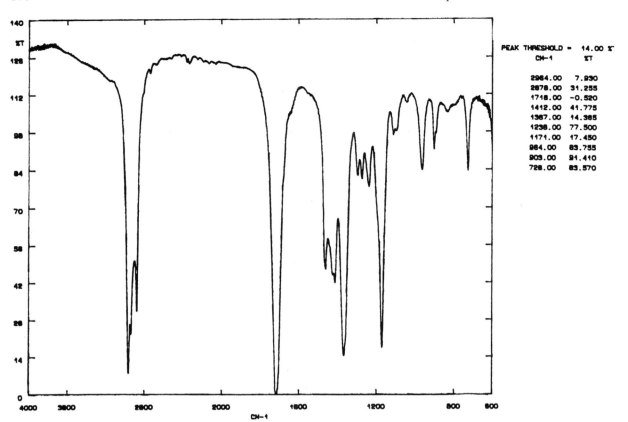

PEAK THRESHOLD = 14.00 %

CM-1	%T
2964.00	7.930
2878.00	31.255
1718.00	-0.520
1412.00	41.775
1367.00	14.385
1238.00	77.500
1171.00	17.450
964.00	83.755
903.00	91.410
726.00	83.570

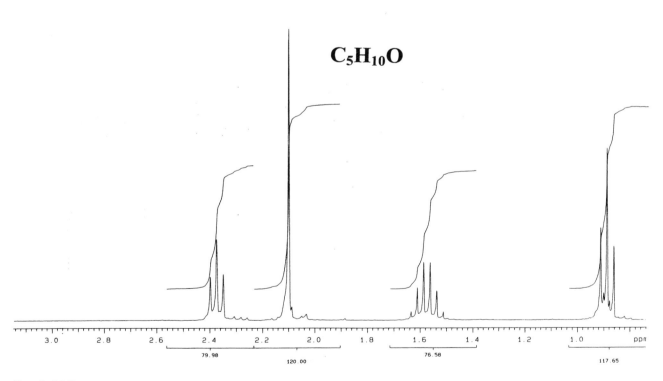

$C_5H_{10}O$

Sample 10.9

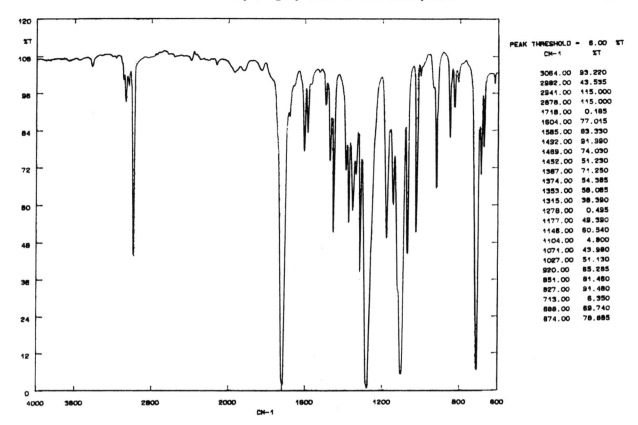

PEAK THRESHOLD = 6.00 %T

CM-1	%T
3064.00	93.220
2982.00	43.535
2941.00	115.000
2878.00	115.000
1718.00	0.185
1604.00	77.015
1585.00	83.330
1492.00	91.990
1469.00	74.030
1452.00	51.230
1387.00	71.250
1374.00	54.385
1353.00	58.085
1315.00	38.390
1278.00	0.495
1177.00	49.390
1146.00	60.540
1104.00	4.900
1071.00	43.990
1027.00	51.130
920.00	65.285
851.00	81.460
827.00	91.480
713.00	6.350
688.00	89.740
674.00	78.885

$C_{10}H_{12}O_2$

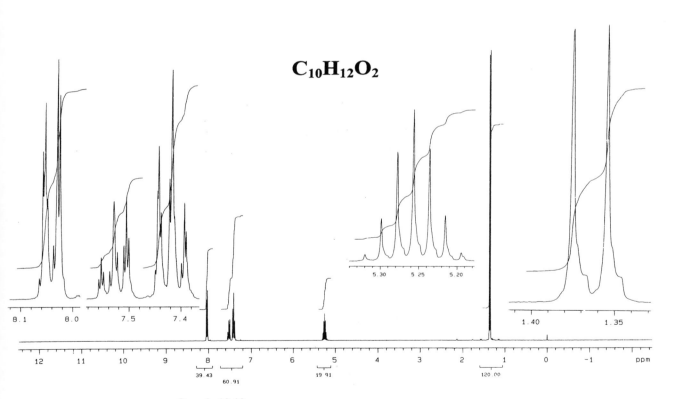

Sample 10.10

Problems in Interpreting Mass Spectra and ^{1}H Nuclear Magnetic Resonance Spectra

When we combine mass spectrometry with ^{1}H NMR spectroscopy we no longer need to supply molecular formulae. It is often a problem to determine the number of hydrogen atoms in a molecule when the integration is uncertain, and it may be best to try to sort out the groups in the molecule and then see how close this comes to the molecular weight. Although ^{1}H NMR is an excellent method for determining structures, the mass spectra must not be ignored—observation, for example, of a McLafferty rearrangement gives a great deal of structural information about a molecule. The mass spectrum can also reveal information about the presence of halogen atoms, and the molecular weight, when compared to the ^{1}H NMR data, can reveal the presence of a plane of symmetry in the molecule.

In this exercise, the mass spectra and ^{1}H NMR spectra of 10 samples are provided. You should be able to identify all the samples from the data.

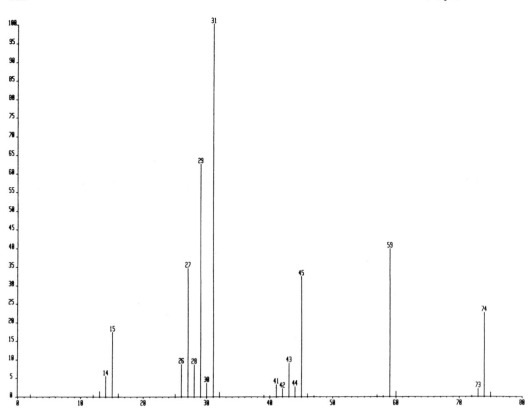

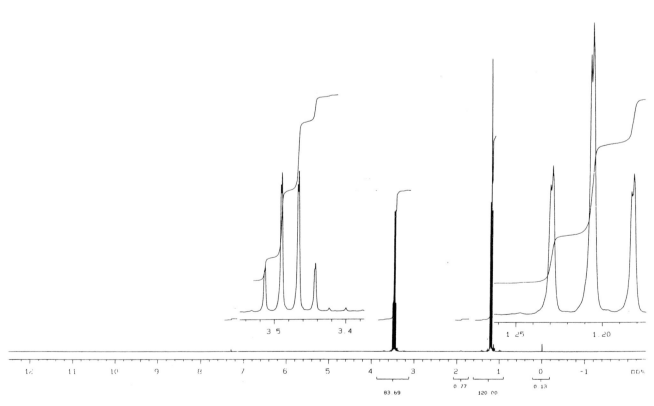

Sample 11.1

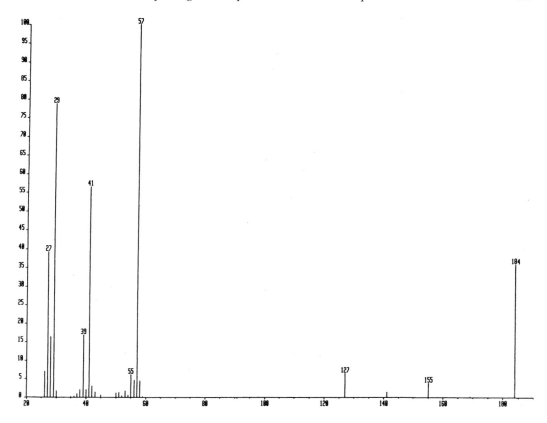

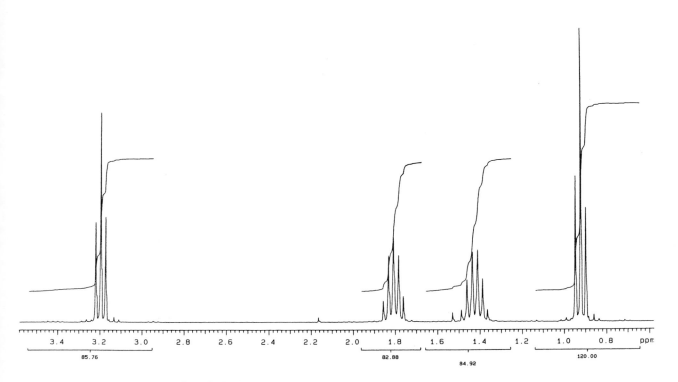

Sample 11.2

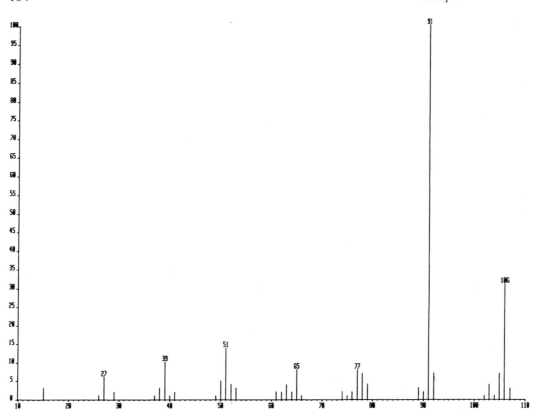

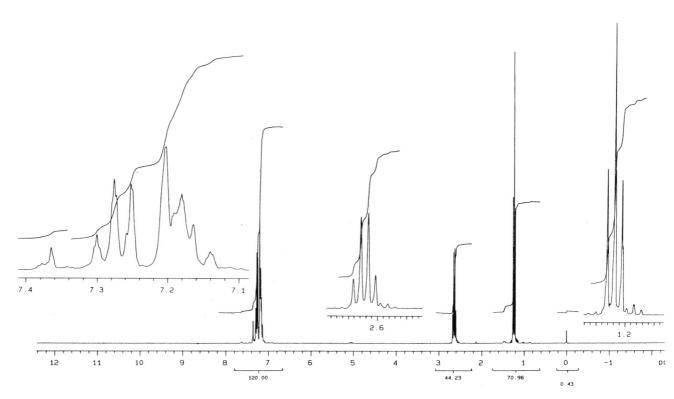

Sample 11.3

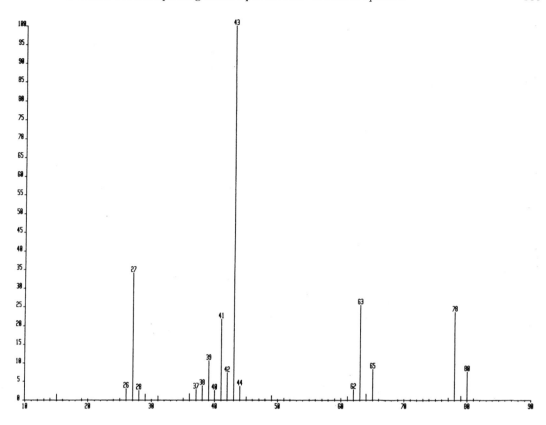

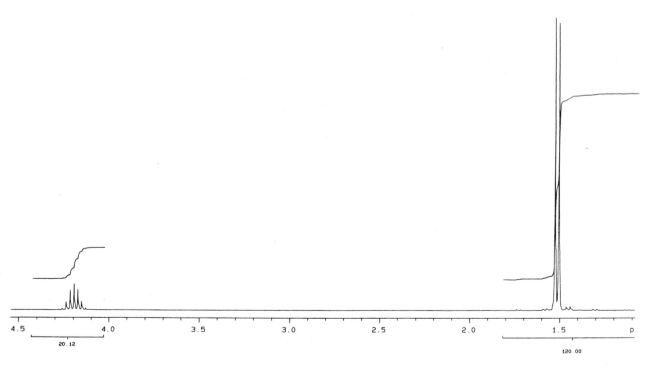

Sample 11.4

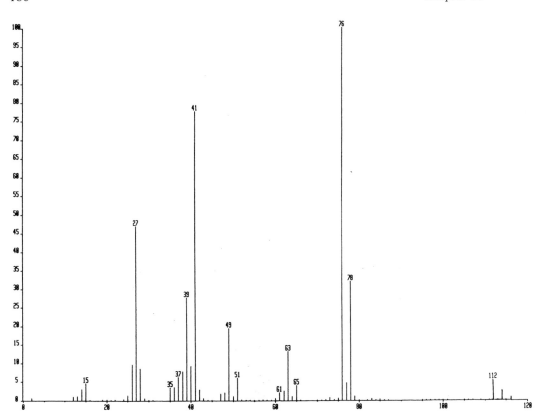

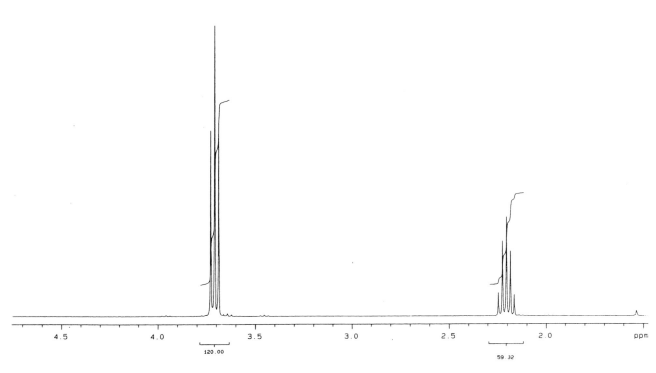

Sample 11.5

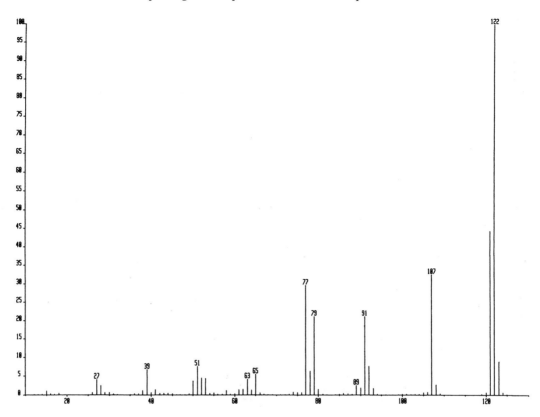

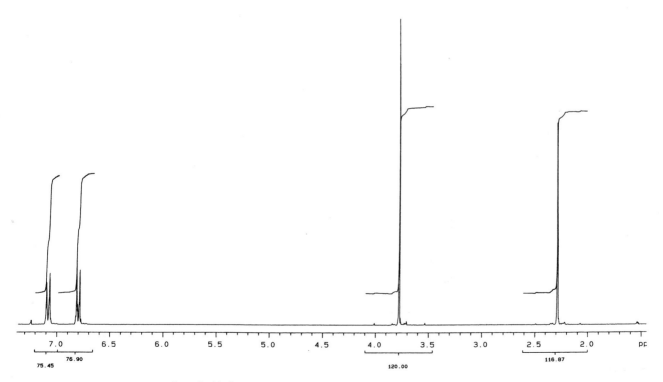

Sample 11.6

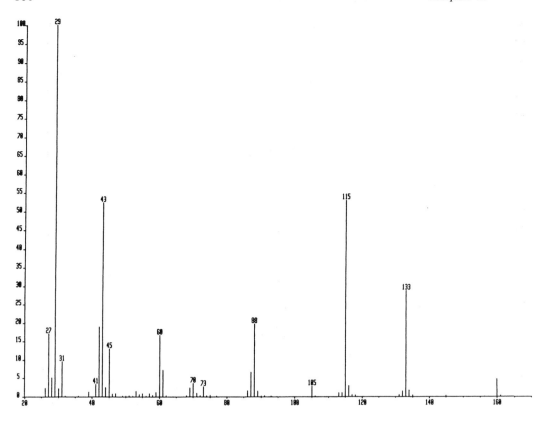

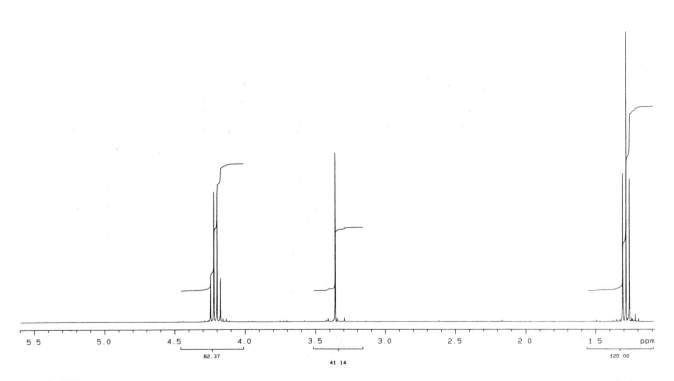

Sample 11.7

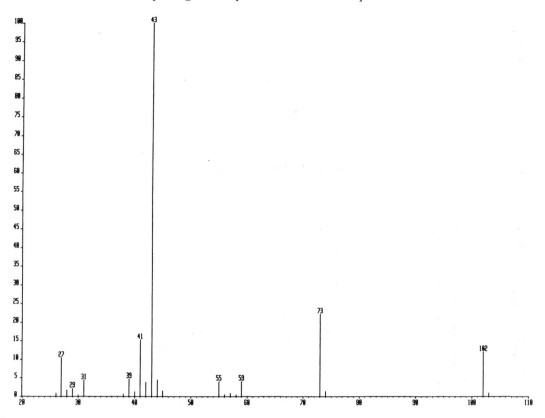

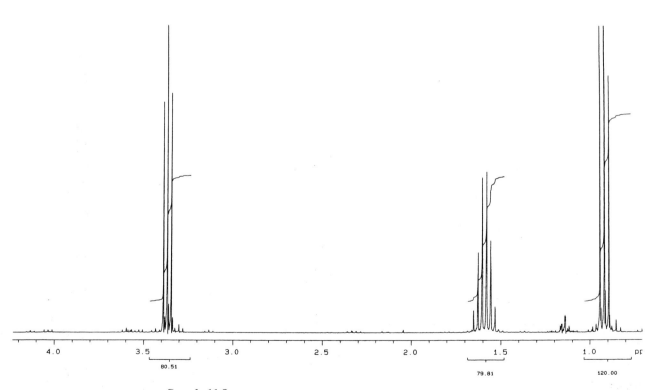

Sample 11.8

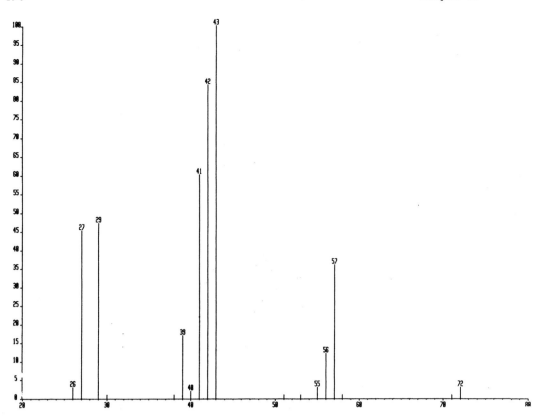

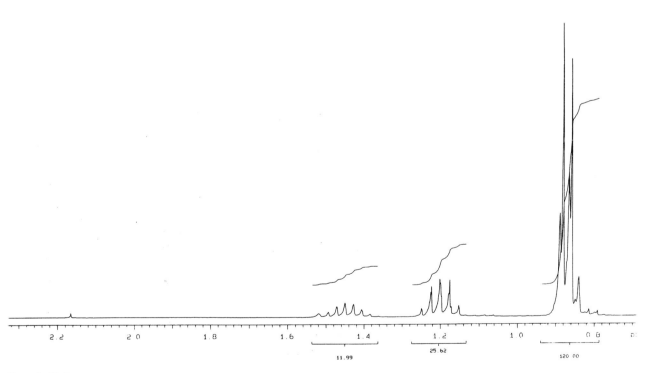

Sample 11.9

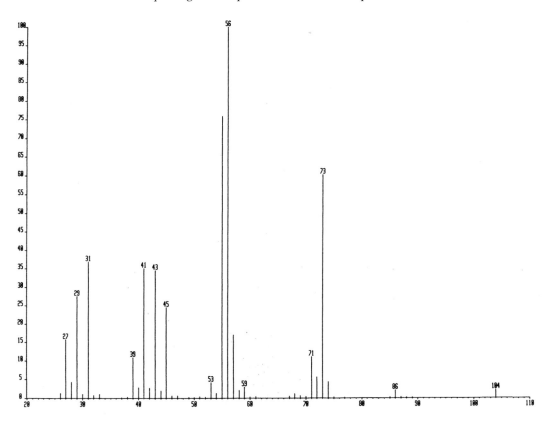

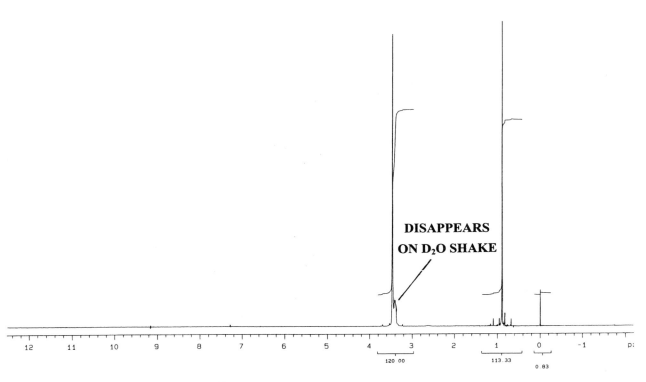

DISAPPEARS
ON D₂O SHAKE

120 00   113.33

0 83

Sample 11.10

Problems in Interpreting Infrared Spectra, Mass Spectra, Ultraviolet Spectra, ^{13}C Nuclear Magnetic Resonance Spectra and ^{1}H Nuclear Magnetic Resonance Spectra

In this chapter we bring all the techniques together, which gives us the ability to identify thousands of simple molecules. However, it also brings the problem of putting together data from up to five techniques to identify a molecule. It is best to start with the infrared spectrum, and to determine the functional groups. Next, the mass spectrum will identify the halogens, reveal the molecular weight and often give us useful structural information. The ultraviolet spectrum can tell us about conjugation in a molecule. In many cases, we can then identify the molecule from the ^{13}C NMR spectrum, but then we must always check that our solution is consistent with the ^{1}H NMR spectrum. In other cases, the ^{13}C NMR spectrum will not provide sufficient information to give complete identification, and the ^{1}H NMR is vital. It is necessary to look back at data obtained—for example, if you missed a fully substituted double bond in the infrared, which is easy to do as the peak is very small, then it would be revealed by the ^{13}C spectrum; you may then have to modify your views on the significance of a carbonyl peak frequency.

It is often difficult to identify alicyclic molecules by ^{1}H NMR, as the spectra become very complicated in rigid systems, but these molecules are relatively easy to identify by ^{13}C NMR. Always consider the symmetry of a molecule—it can often be used to distinguish between possibilities, particularly when considering ^{13}C spectra.

This exercise is the final test of your ability to bring together the data from all techniques. In each case you are given the infrared, mass, ^{13}C and ^{1}H NMR spectra and, where appropriate, the ultraviolet spectra. You should be able to identify all the samples from the data provided.

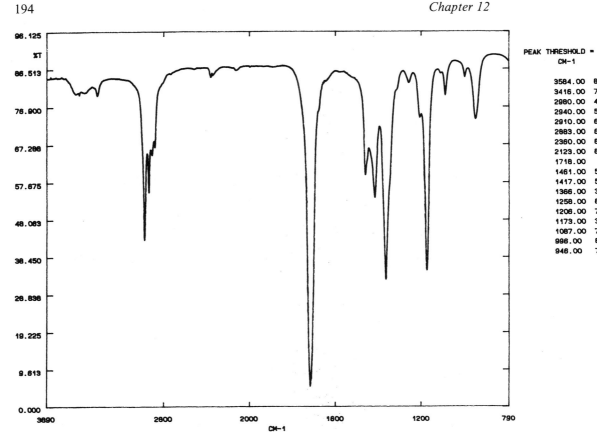

PEAK THRESHOLD = 0.96 %T
CM-1   %T

3584.00   80.210
3416.00   79.970
2980.00   43.075
2940.00   55.345
2910.00   64.900
2883.00   66.895
2360.00   84.670
2123.00   86.415
1718.00   5.410
1461.00   59.675
1417.00   53.885
1366.00   32.830
1258.00   83.160
1206.00   74.245
1173.00   35.205
1087.00   79.970
996.00   84.585
946.00   73.845

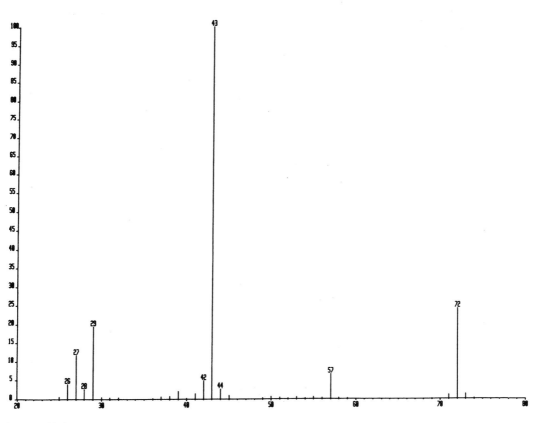

Sample 12.1

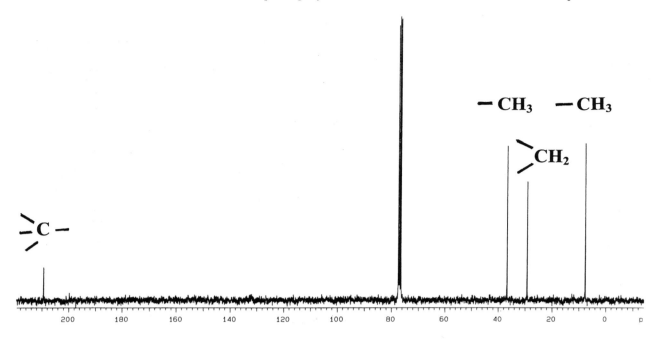

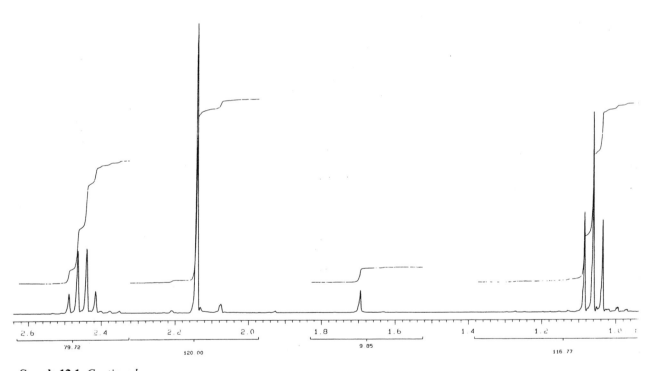

Sample 12.1 *Continued*

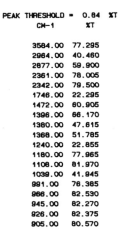

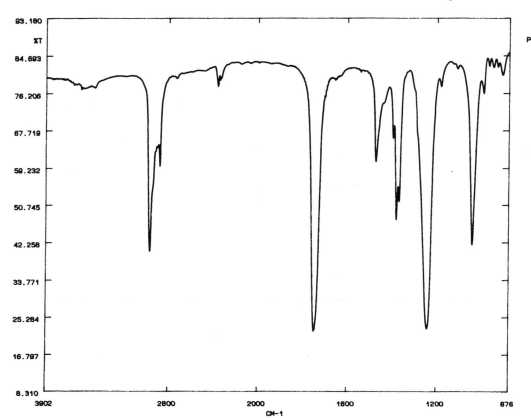

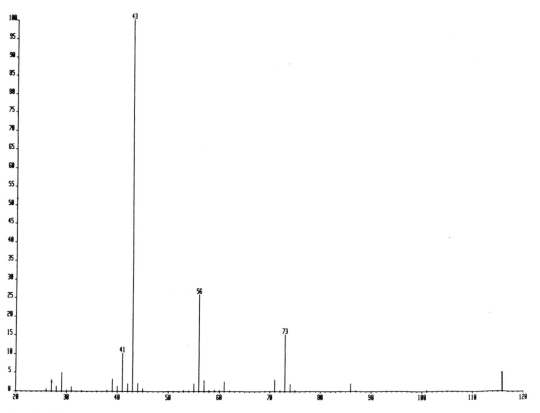

Sample 12.2

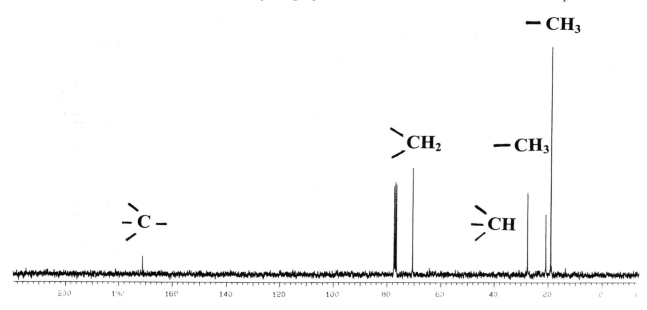

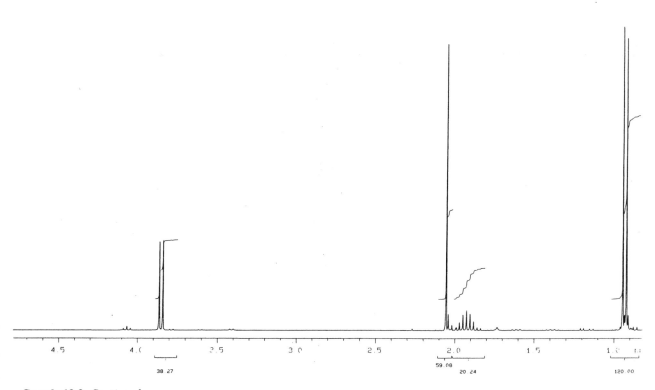

Sample 12.2 *Continued*

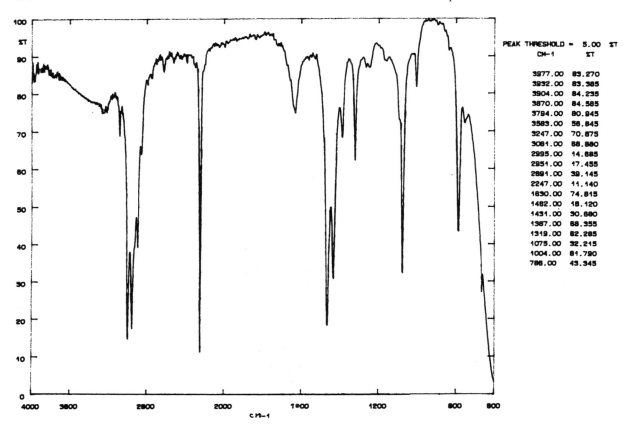

PEAK THRESHOLD = 5.00 %T

CM-1	%T
3977.00	83.270
3932.00	83.385
3904.00	84.235
3870.00	84.585
3794.00	80.945
3583.00	58.845
3247.00	70.875
3081.00	88.880
2995.00	14.885
2951.00	17.455
2891.00	39.145
2247.00	11.140
1830.00	74.815
1482.00	18.120
1431.00	30.880
1387.00	88.355
1319.00	82.285
1075.00	32.215
1004.00	81.790
786.00	43.345

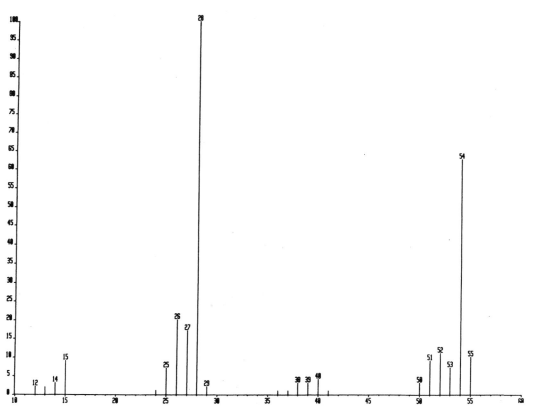

Sample 12.3

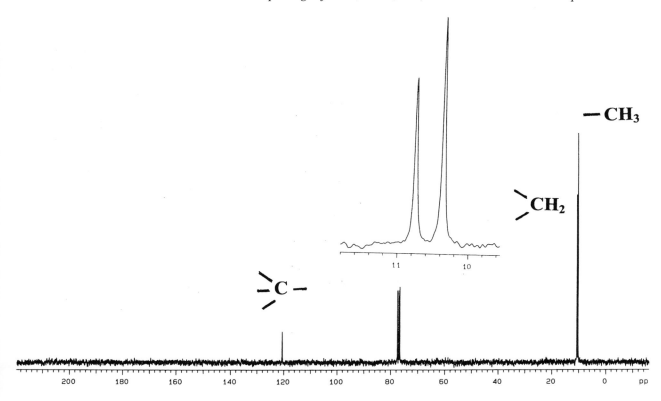

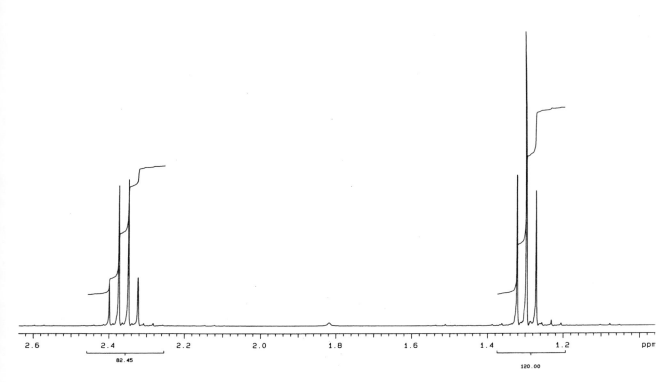

Sample 12.3 *Continued*

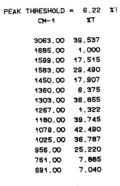

PEAK THRESHOLD =   6.22   %T
      CM-1   %T

    3063.00   39.537
    1685.00   1.000
    1599.00   17.515
    1583.00   29.490
    1450.00   17.907
    1360.00   6.375
    1303.00   36.855
    1267.00   1.322
    1180.00   39.745
    1079.00   42.490
    1025.00   36.787
     956.00   25.220
     761.00   7.885
     691.00   7.040

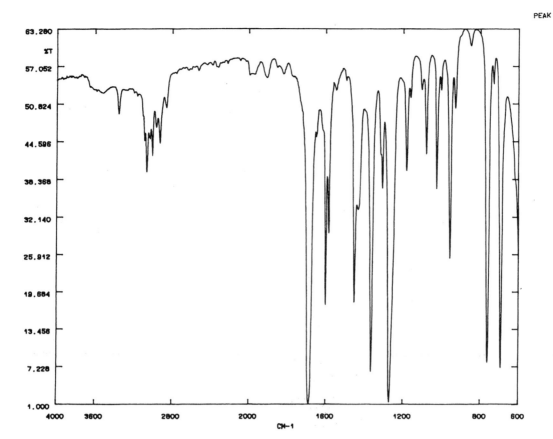

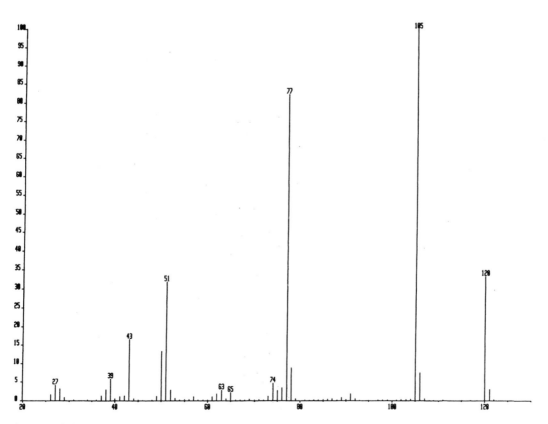

Sample 12.4

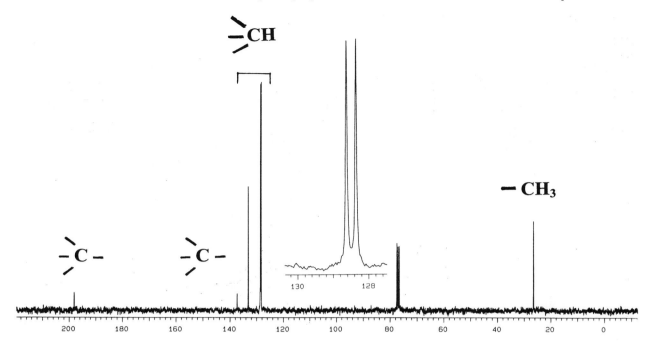

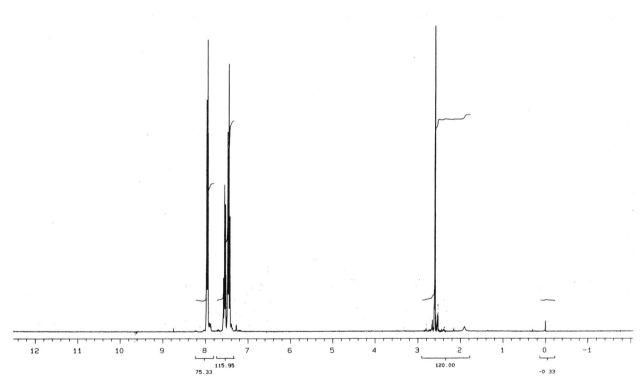

Sample 12.4 *Continued*

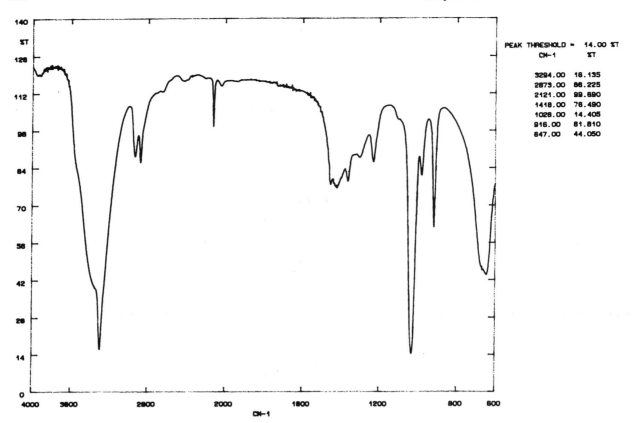

PEAK THRESHOLD = 14.00 %T

CM-1	%T
3294.00	16.135
2873.00	86.225
2121.00	99.690
1418.00	76.490
1028.00	14.405
916.00	61.810
847.00	44.050

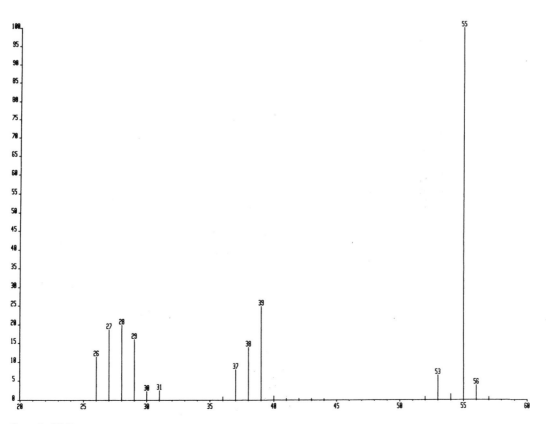

Sample 12.5

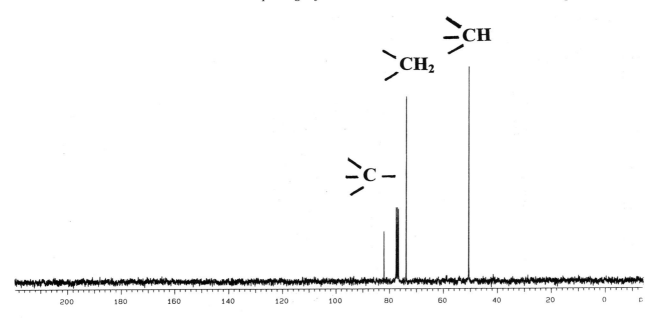

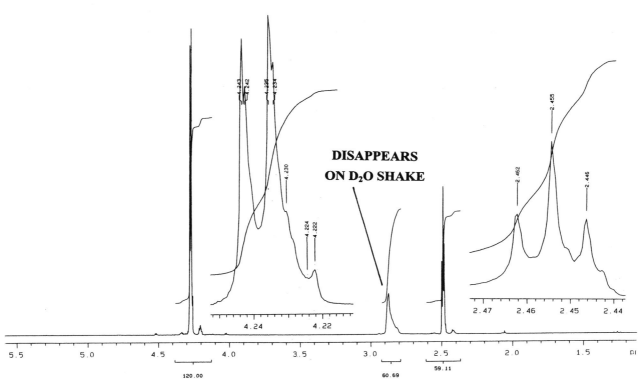

Sample 12.5 *Continued*

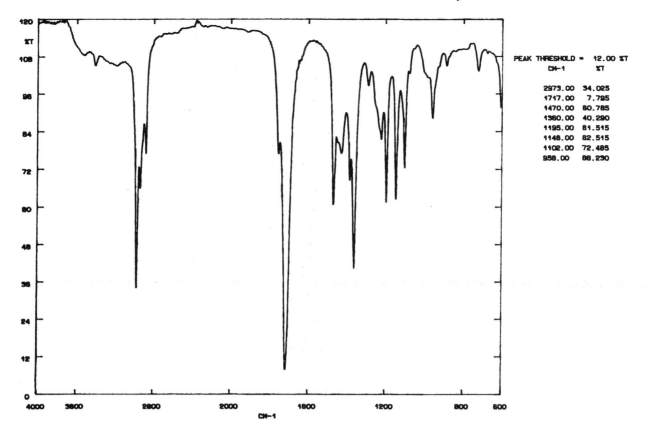

PEAK THRESHOLD =	12.00 %T
CM-1	%T
2973.00	34.025
1717.00	7.795
1470.00	60.785
1360.00	40.290
1195.00	61.515
1146.00	62.515
1102.00	72.485
958.00	88.230

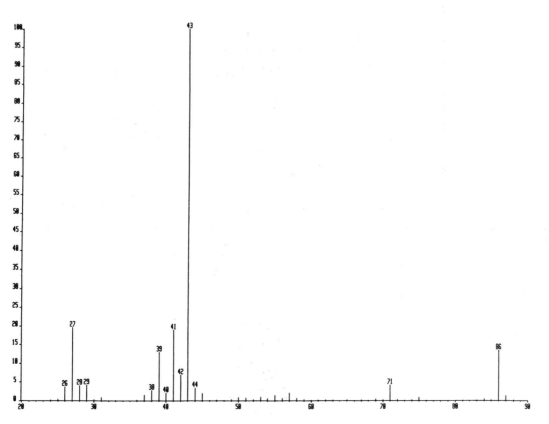

Sample 12.6

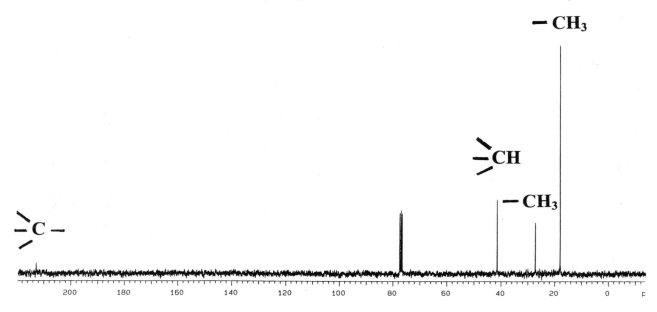

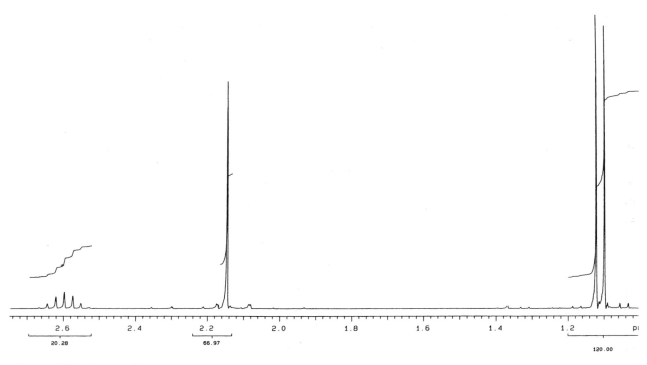

Sample 12.6 *Continued*

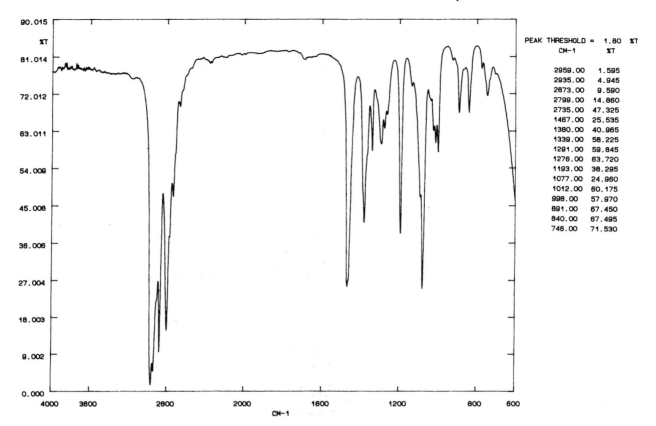

PEAK THRESHOLD =   1.80 %T

CM-1	%T
2959.00	1.595
2935.00	4.945
2873.00	9.590
2799.00	14.860
2735.00	47.325
1467.00	25.535
1380.00	40.965
1339.00	58.225
1291.00	59.845
1276.00	63.720
1193.00	38.295
1077.00	24.960
1012.00	60.175
998.00	57.970
891.00	67.450
840.00	67.495
746.00	71.530

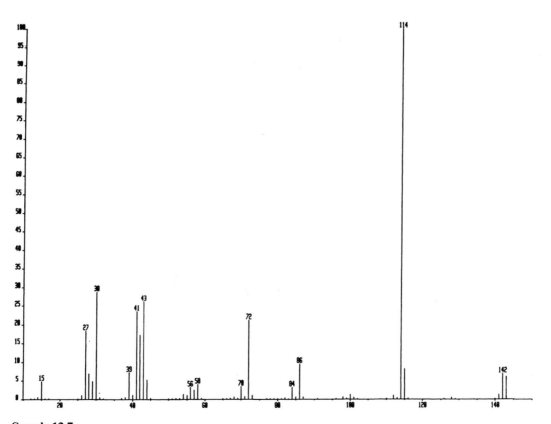

Sample 12.7

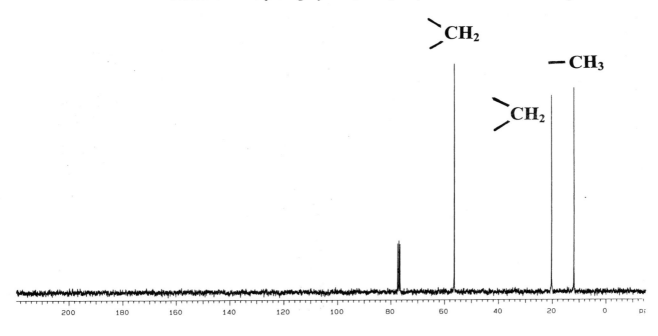

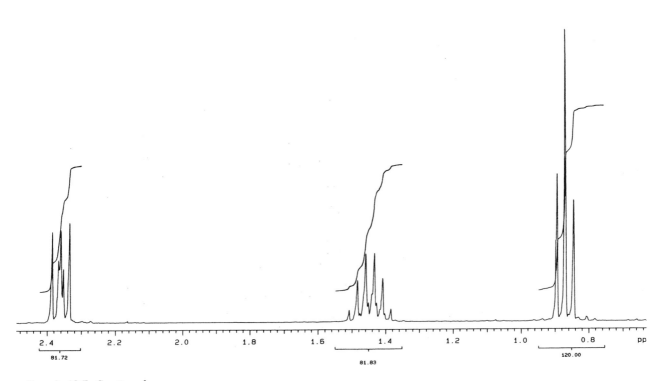

Sample 12.7 *Continued*

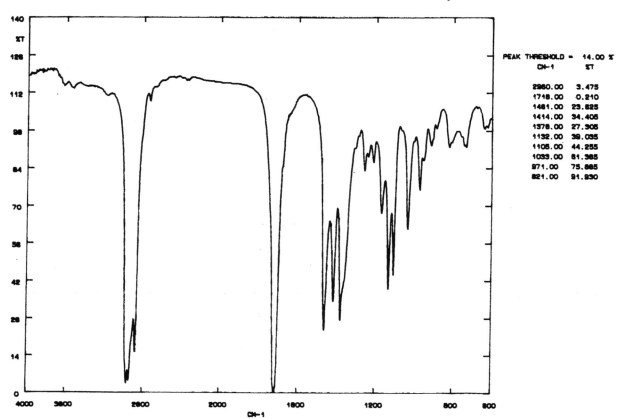

PEAK THRESHOLD = 14.00 %

CM-1	%T
2960.00	3.475
1718.00	0.210
1461.00	23.625
1414.00	34.405
1378.00	27.905
1132.00	39.095
1105.00	44.255
1033.00	61.365
971.00	75.865
821.00	91.930

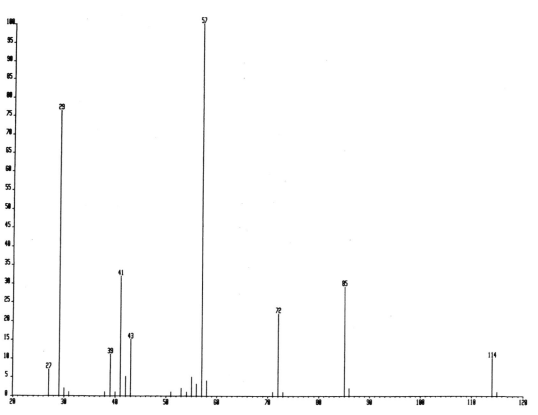

Sample 12.8

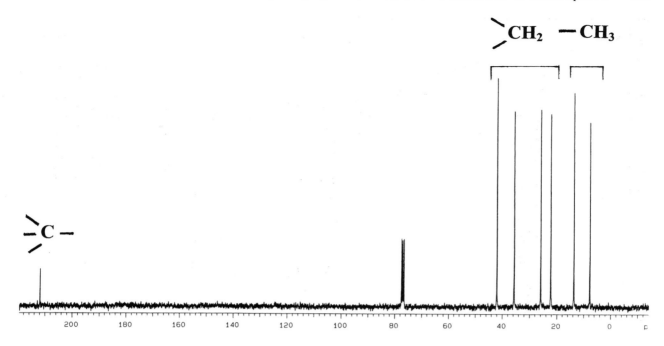

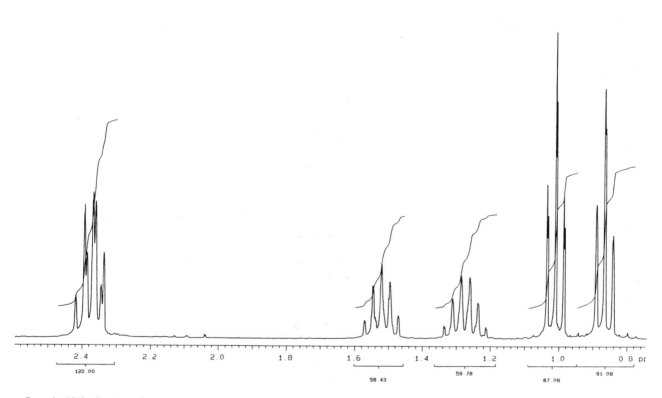

Sample 12.8 *Continued*

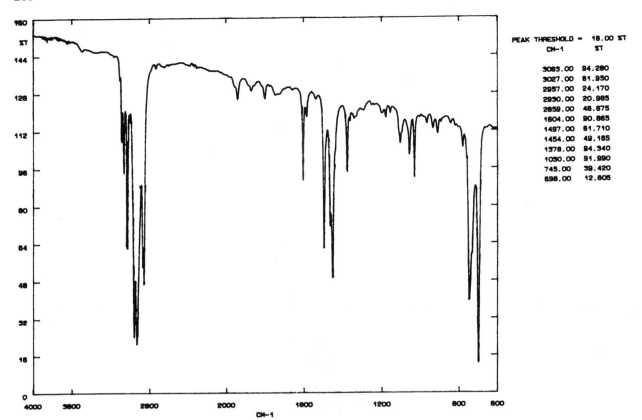

PEAK THRESHOLD =   18.00 %T
       CM-1   %T

     3083.00   94.280
     3027.00   81.950
     2957.00   24.170
     2930.00   20.985
     2859.00   48.675
     1604.00   90.865
     1497.00   81.710
     1454.00   49.165
     1378.00   94.340
     1090.00   91.990
      745.00   39.420
      698.00   12.605

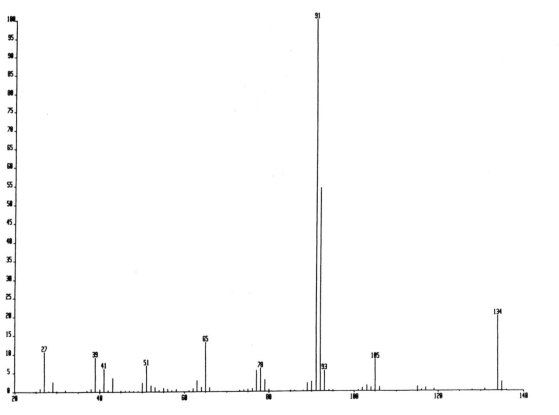

Sample 12.9

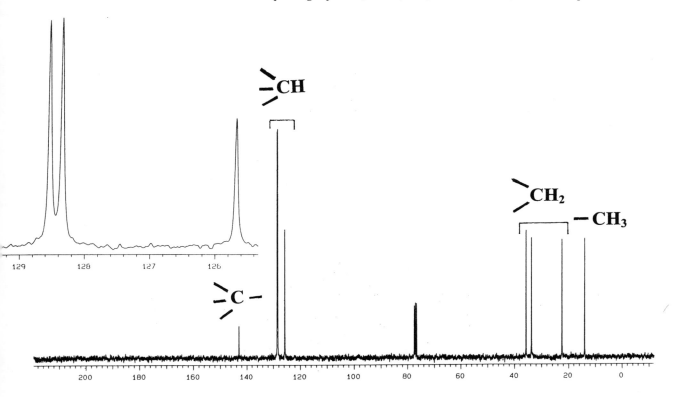

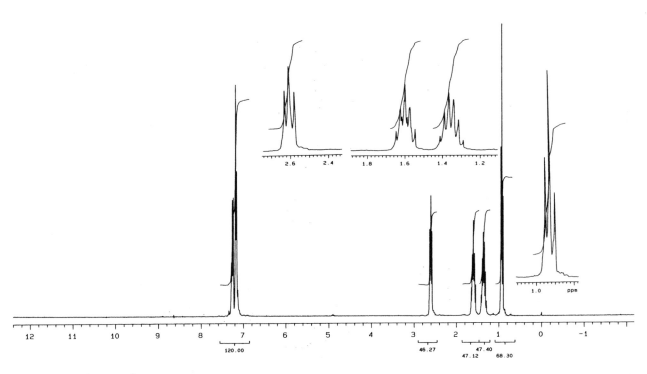

Sample 12.9 *Continued*

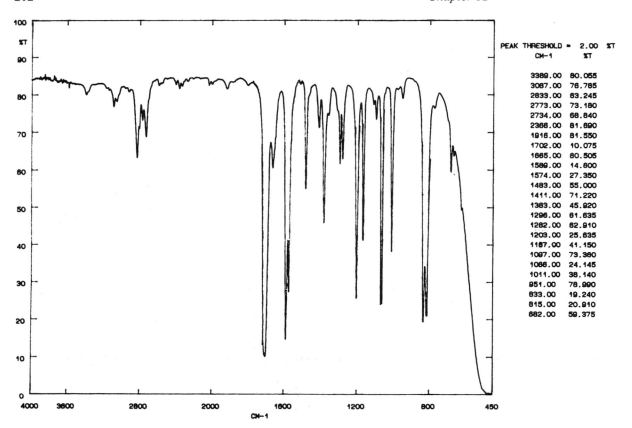

PEAK THRESHOLD = 2.00 %T

CM-1	%T
3389.00	80.055
3087.00	78.785
2833.00	63.245
2773.00	73.180
2734.00	68.840
2366.00	81.690
1916.00	81.550
1702.00	10.075
1665.00	80.505
1589.00	14.600
1574.00	27.350
1483.00	55.000
1411.00	71.220
1363.00	45.920
1296.00	61.635
1282.00	62.910
1203.00	25.635
1157.00	41.150
1097.00	73.380
1066.00	24.145
1011.00	38.140
951.00	78.990
833.00	19.240
815.00	20.910
682.00	59.375

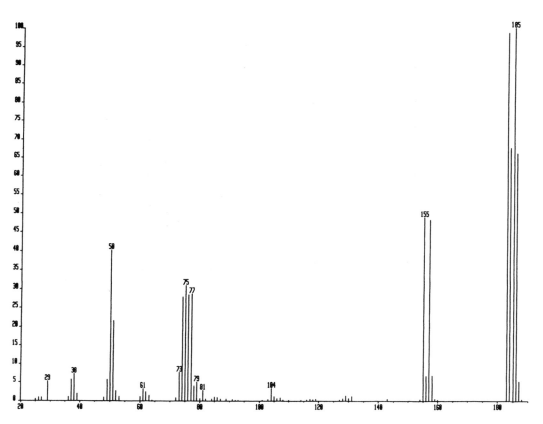

Sample 12.10

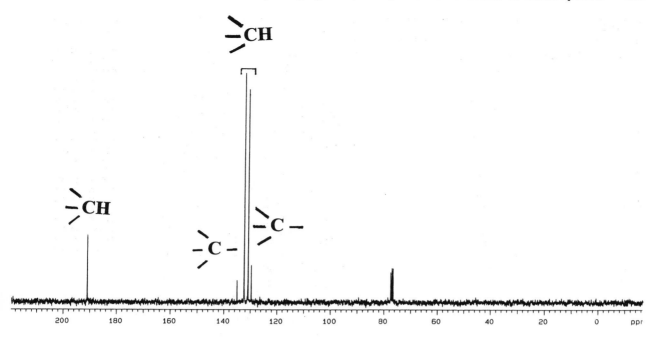

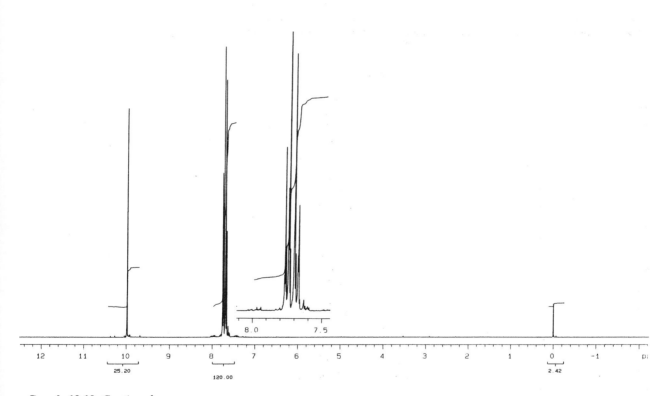

Sample 12.10 *Continued*

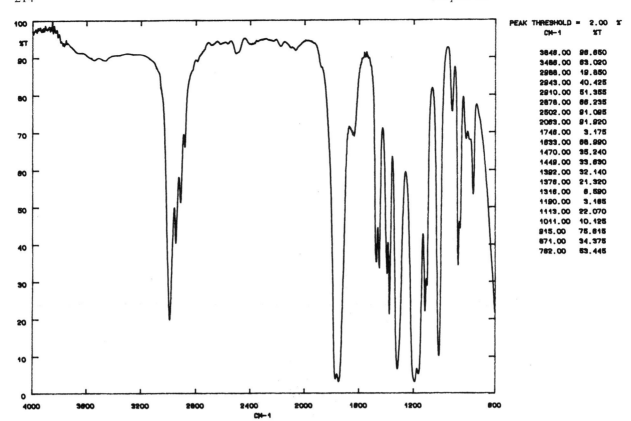

PEAK THRESHOLD = 2.00 %

CM-1	%T
3846.00	96.650
3466.00	63.020
2966.00	19.850
2943.00	40.425
2910.00	51.355
2878.00	66.235
2502.00	91.095
2063.00	91.920
1746.00	3.175
1633.00	66.990
1470.00	35.240
1449.00	33.630
1392.00	32.140
1376.00	21.320
1316.00	6.590
1190.00	3.185
1113.00	22.070
1011.00	10.125
915.00	75.615
871.00	34.375
762.00	63.445

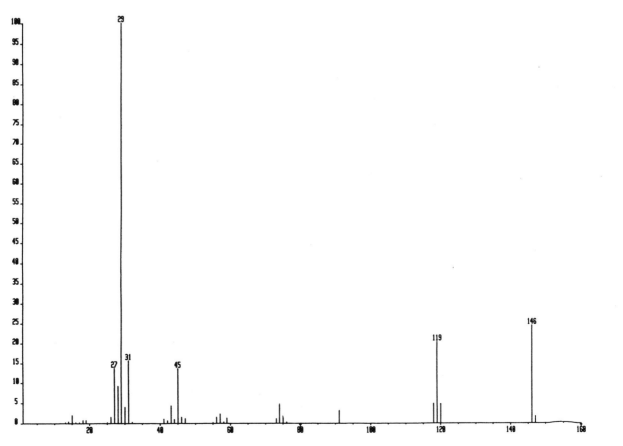

Sample 12.11

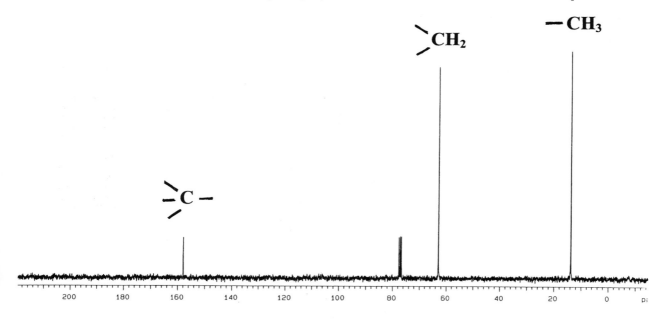

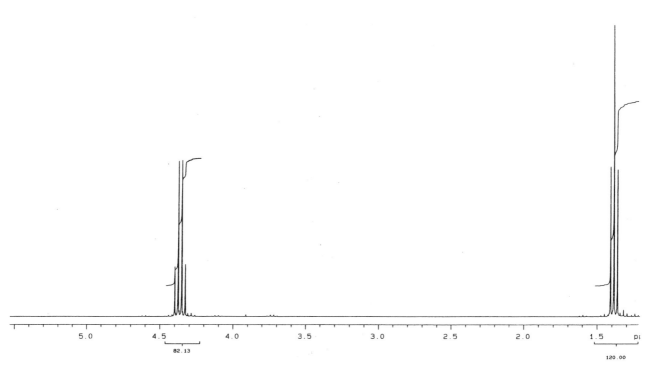

Sample 12.11 *Continued*

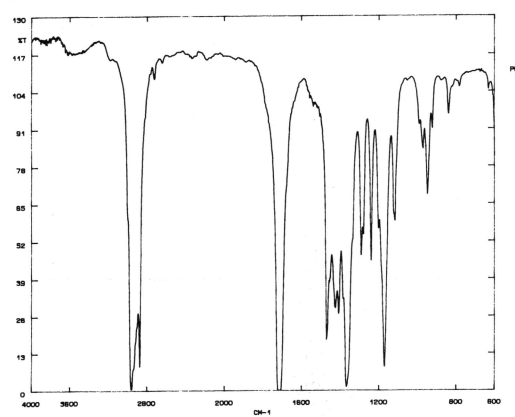

PEAK THRESHOLD =   13.00 %T

CM-1   %T

3416.00   107.090
2959.00   0.325
2873.00   8.420
1718.00   0.875
1470.00   17.750
1408.00   26.835
1387.00   1.175
1291.00   47.100
1240.00   45.205
1171.00   8.200
1118.00   58.885
949.00   68.190

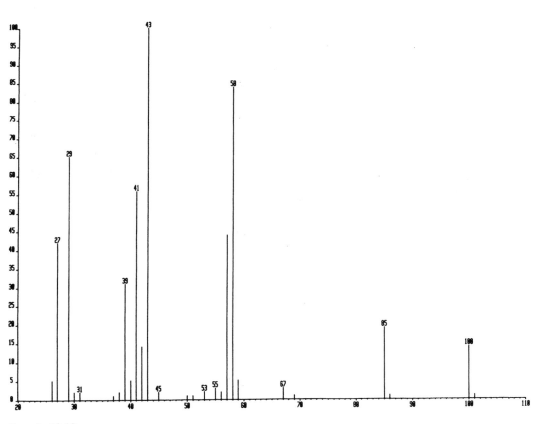

Sample 12.12

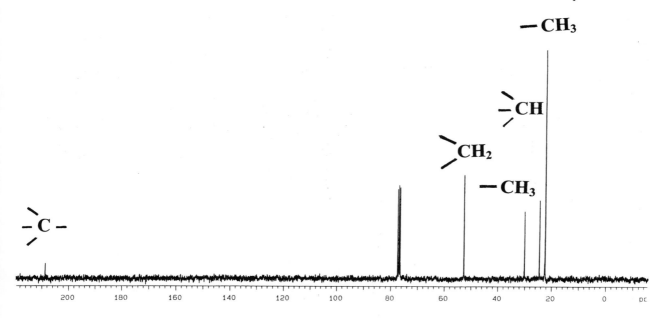

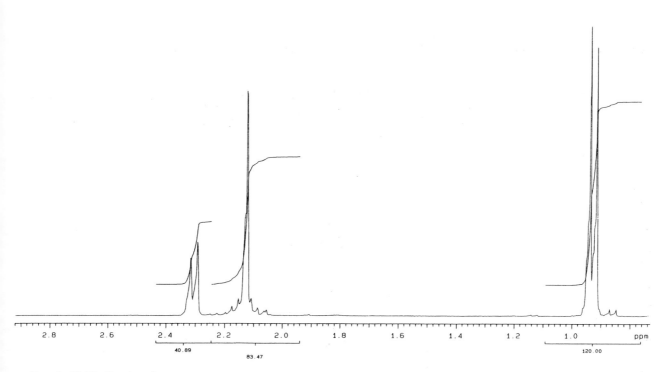

Sample 12.12 *Continued*

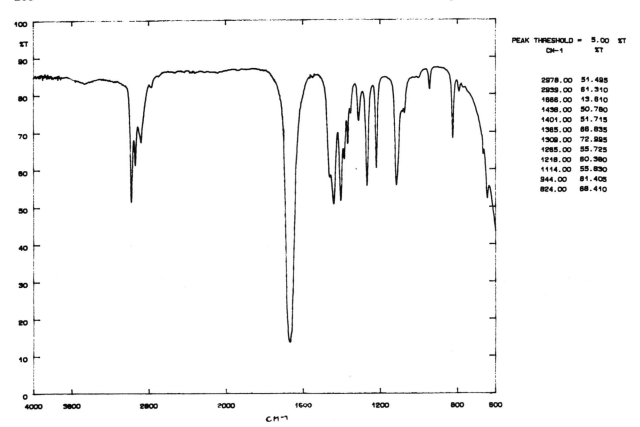

PEAK THRESHOLD = 5.00 %T

CM-1	%T
2978.00	51.495
2939.00	61.310
1666.00	13.810
1436.00	50.780
1401.00	51.715
1365.00	66.835
1309.00	72.995
1265.00	55.725
1218.00	60.380
1114.00	55.830
944.00	61.405
824.00	66.410

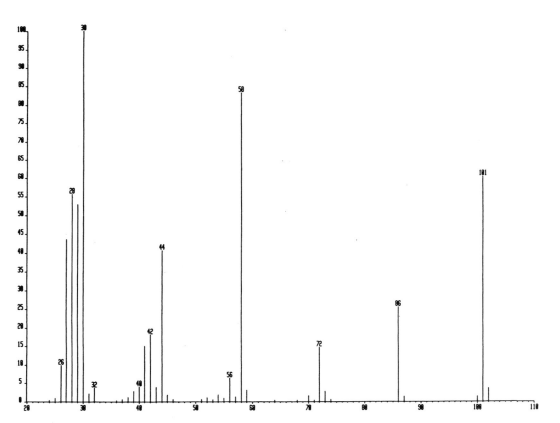

Sample 12.13

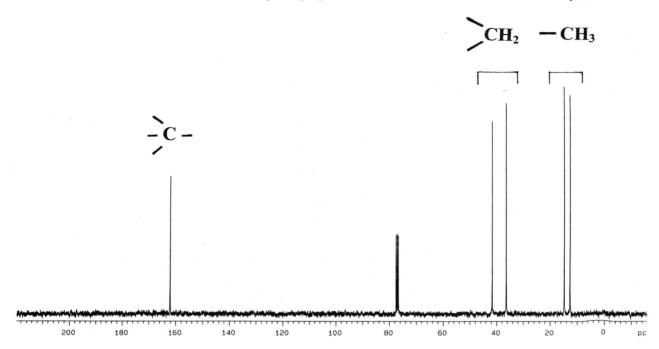

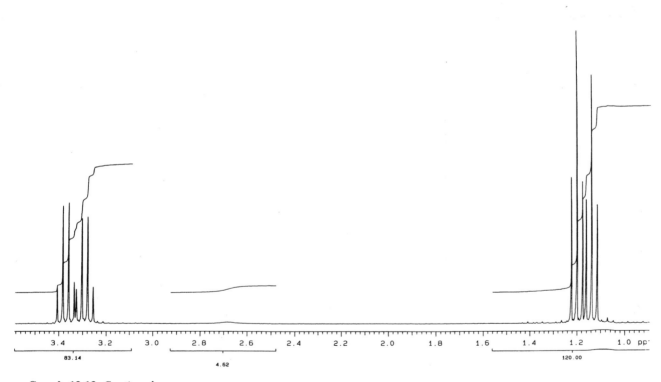

Sample 12.13 *Continued*

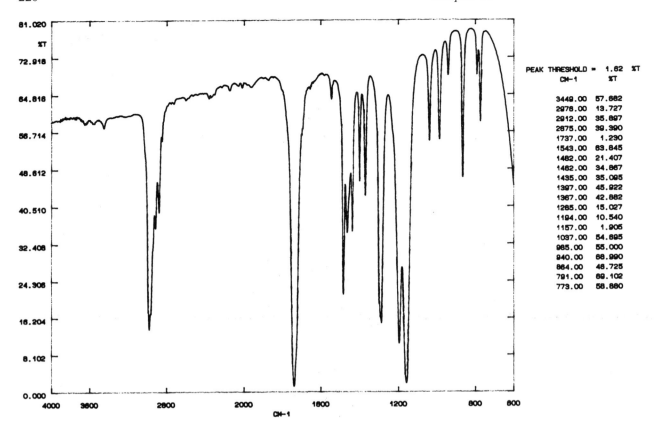

PEAK THRESHOLD = 1.62 %T

CM-1	%T
3449.00	57.682
2976.00	13.727
2912.00	35.897
2875.00	39.390
1737.00	1.290
1543.00	63.845
1482.00	21.407
1462.00	34.867
1435.00	35.095
1397.00	45.922
1367.00	42.882
1265.00	15.027
1194.00	10.540
1157.00	1.905
1037.00	54.695
985.00	55.000
940.00	68.990
864.00	48.725
791.00	69.102
773.00	58.880

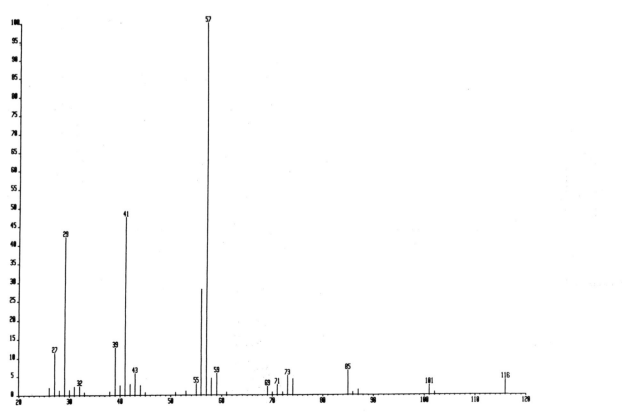

Sample 12.14

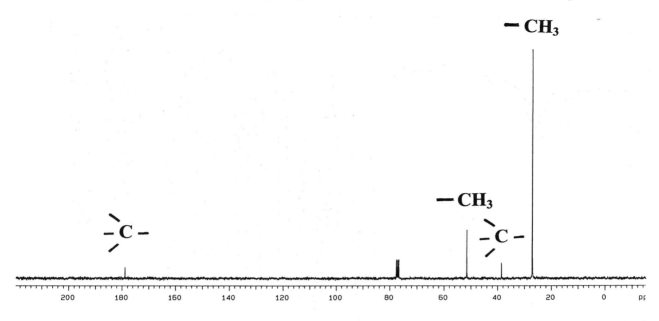

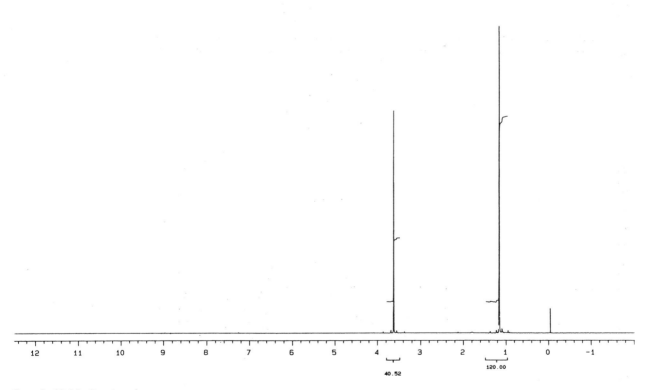

Sample 12.14 *Continued*

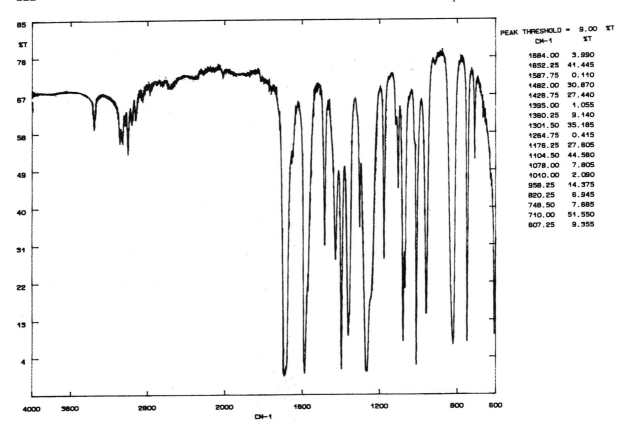

PEAK THRESHOLD = 9.00 %T

CM-1	%T
1684.00	3.990
1652.25	41.445
1587.75	0.110
1482.00	30.870
1428.75	27.440
1395.00	1.055
1360.25	9.140
1301.50	35.185
1264.75	0.415
1176.25	27.805
1104.50	44.580
1078.00	7.805
1010.00	2.090
958.25	14.375
820.25	6.945
748.50	7.685
710.00	51.550
607.25	9.355

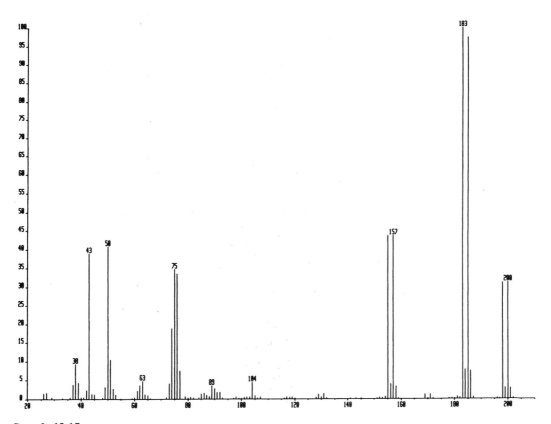

Sample 12.15

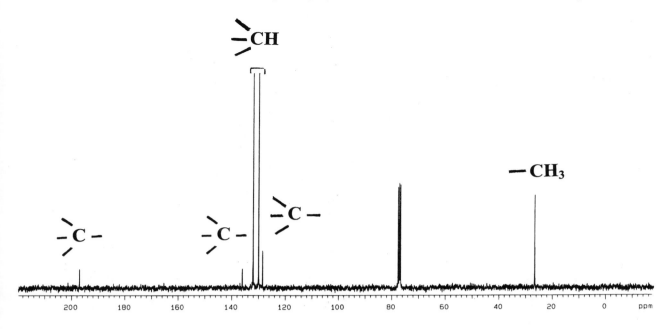

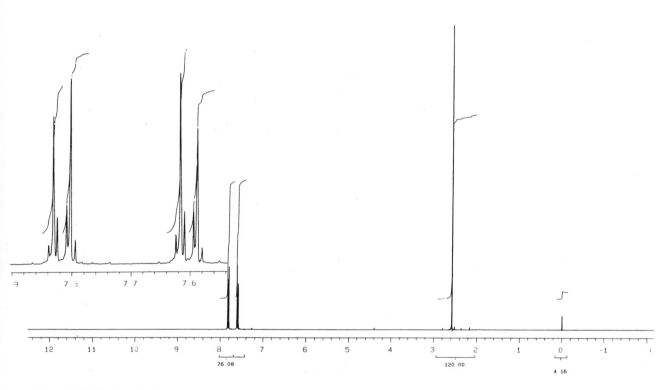

Sample 12.15 *Continued*

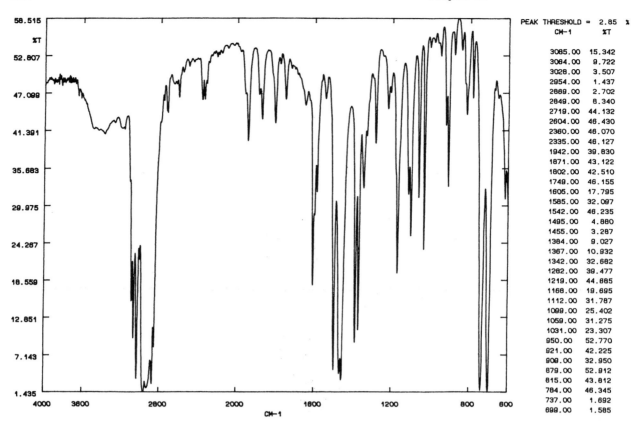

PEAK THRESHOLD = 2.85 %
CM-1	%T
3085.00	15.342
3064.00	9.722
3028.00	3.507
2954.00	1.437
2869.00	2.702
2849.00	8.340
2719.00	44.132
2604.00	46.430
2360.00	46.070
2335.00	46.127
1942.00	39.830
1871.00	43.122
1802.00	42.510
1749.00	46.155
1605.00	17.795
1585.00	32.097
1542.00	46.235
1495.00	4.880
1455.00	3.287
1384.00	9.027
1367.00	10.932
1342.00	32.682
1282.00	39.477
1219.00	44.685
1168.00	19.695
1112.00	31.787
1099.00	25.402
1059.00	31.275
1031.00	23.307
950.00	52.770
921.00	42.225
909.00	32.950
879.00	52.912
815.00	43.812
784.00	46.345
737.00	1.892
699.00	1.585

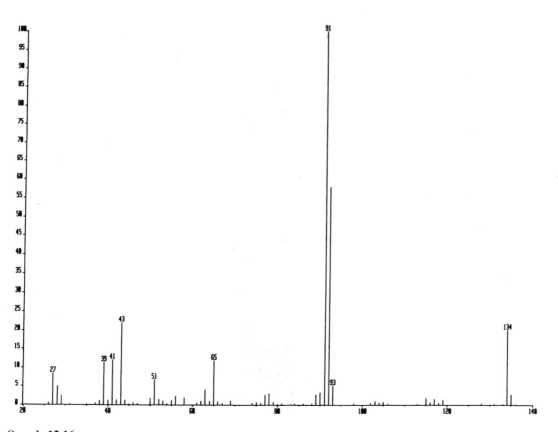

Sample 12.16

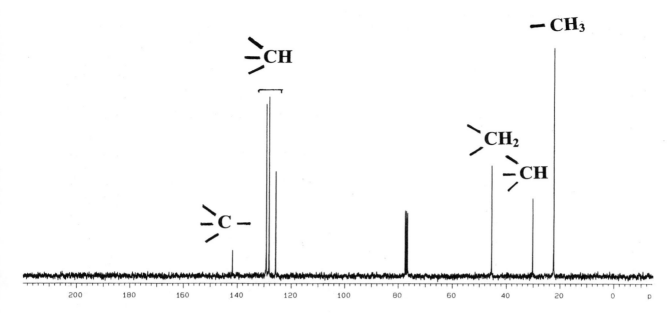

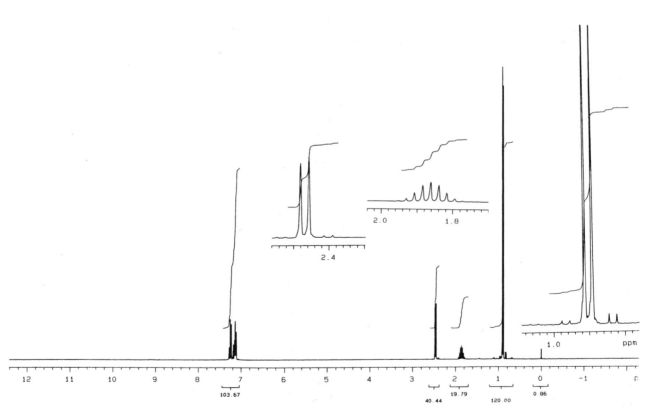

Sample 12.16 *Continued*

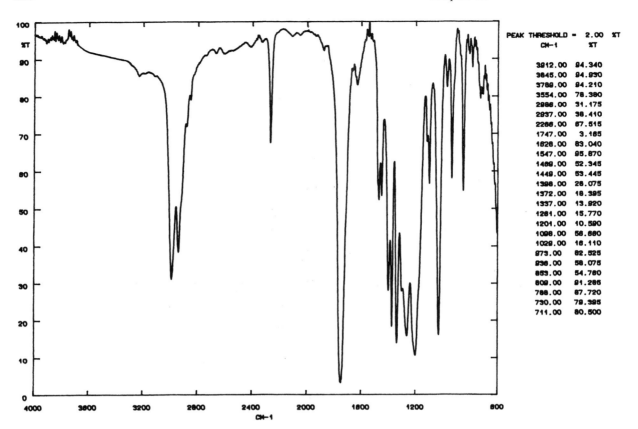

PEAK THRESHOLD = 2.00 %T

CM-1	%T
3912.00	94.340
3845.00	94.930
3789.00	94.210
3554.00	78.380
2986.00	31.175
2937.00	38.410
2266.00	87.515
1747.00	3.185
1626.00	83.040
1547.00	95.870
1469.00	52.345
1449.00	53.445
1398.00	26.075
1372.00	18.395
1337.00	13.920
1261.00	15.770
1201.00	10.590
1098.00	56.660
1029.00	16.110
973.00	82.525
936.00	58.075
853.00	54.780
809.00	91.285
788.00	87.720
730.00	79.395
711.00	80.500

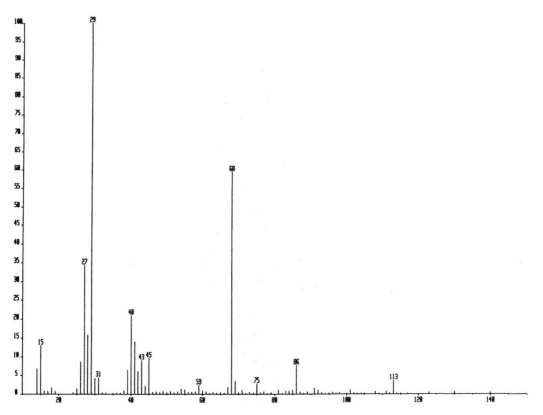

Sample 12.17

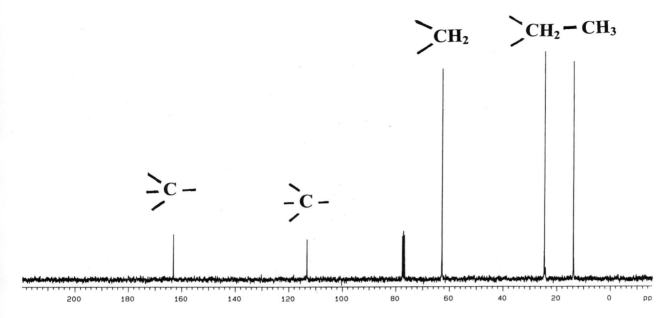

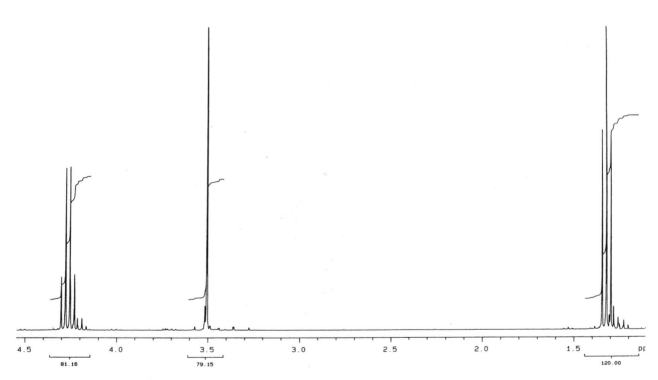

Sample 12.17 *Continued*

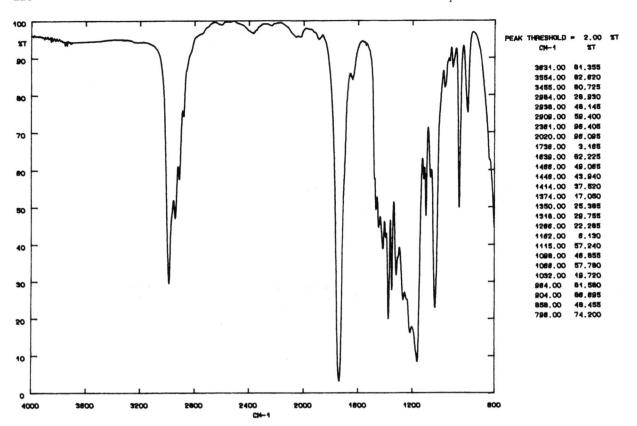

PEAK THRESHOLD = 2.00 %T

CM-1	%T
3631.00	81.355
3554.00	82.620
3455.00	80.725
2964.00	26.930
2938.00	46.145
2909.00	59.400
2361.00	96.405
2020.00	96.095
1736.00	3.165
1639.00	82.225
1466.00	45.065
1446.00	43.940
1414.00	37.520
1374.00	17.050
1350.00	25.365
1318.00	29.755
1266.00	22.265
1162.00	6.130
1115.00	57.240
1098.00	46.855
1066.00	57.780
1032.00	19.720
964.00	81.580
904.00	86.695
858.00	46.455
796.00	74.200

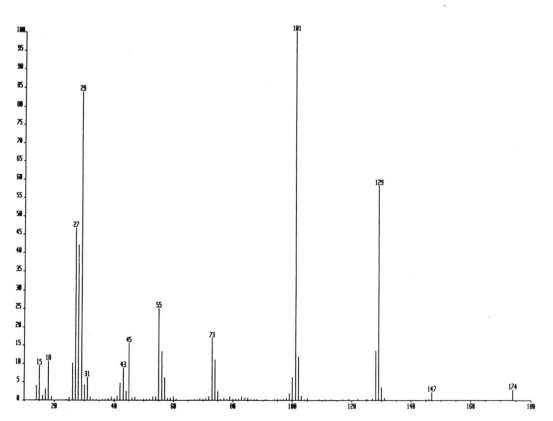

Sample 12.18

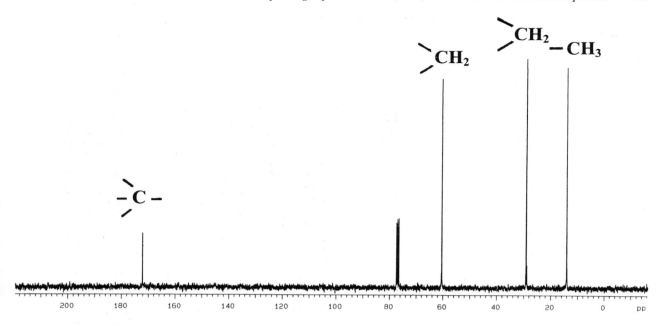

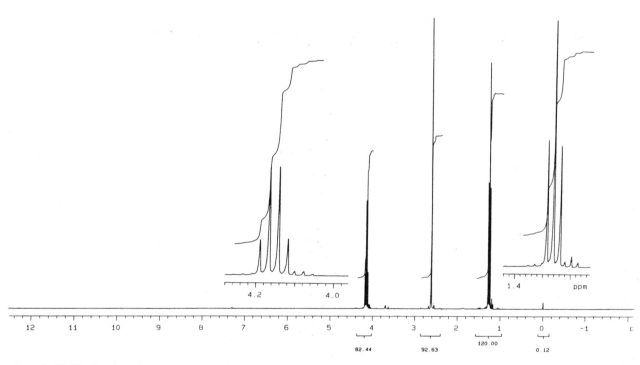

Sample 12.18 *Continued*

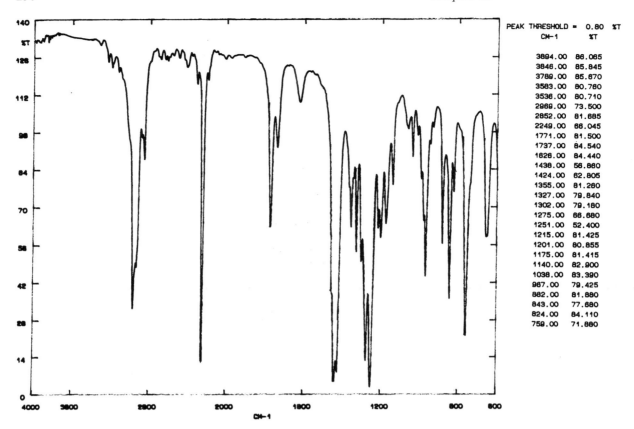

PEAK THRESHOLD = 0.80 %T

CM-1	%T
3894.00	86.065
3846.00	85.845
3789.00	85.670
3583.00	80.760
3536.00	80.710
2969.00	73.500
2852.00	81.685
2249.00	66.045
1771.00	81.500
1737.00	84.540
1626.00	84.440
1436.00	56.860
1424.00	62.805
1355.00	81.260
1327.00	79.840
1302.00	79.180
1275.00	66.680
1251.00	52.400
1215.00	81.425
1201.00	80.855
1175.00	81.415
1140.00	82.900
1038.00	83.390
967.00	79.425
882.00	81.880
843.00	77.680
824.00	84.110
759.00	71.880

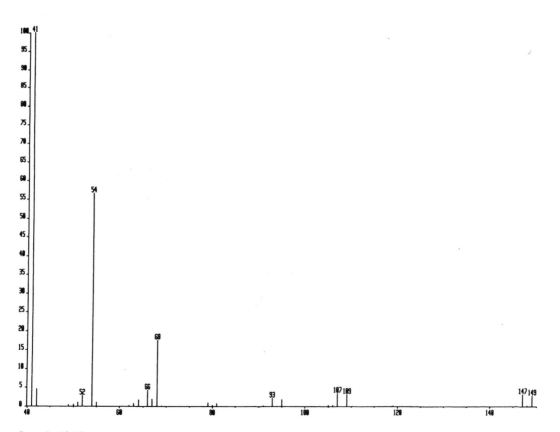

Sample 12.19

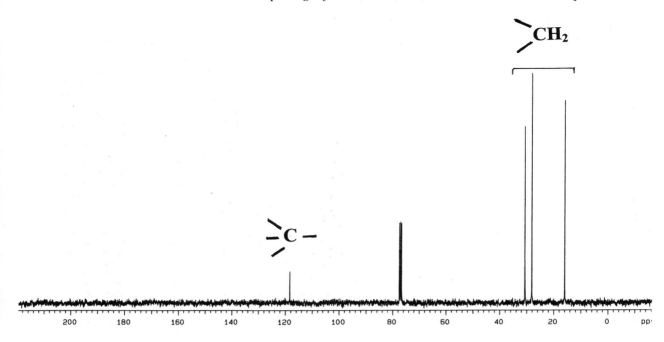

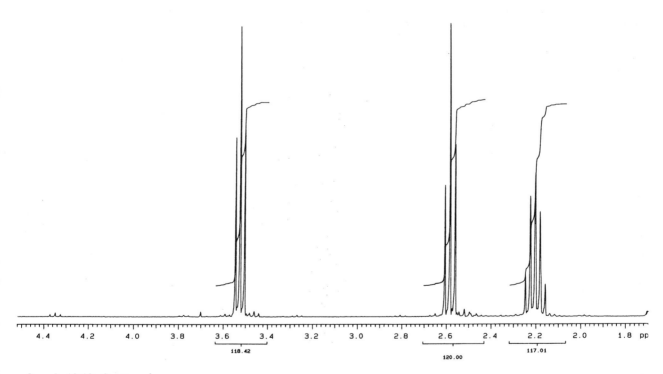

Sample 12.19 *Continued*

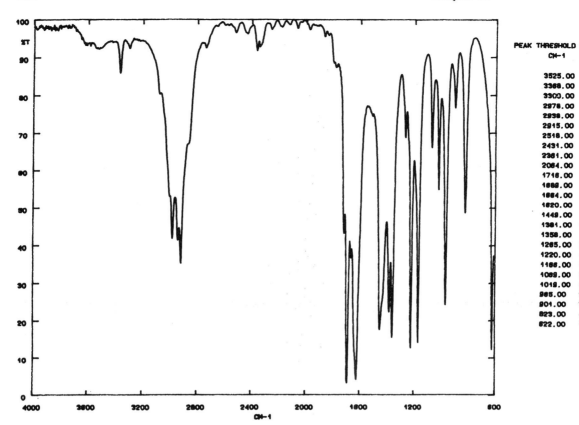

PEAK THRESHOLD = 2.00 %T
CM-1   %T

3525.00   92.175
3368.00   85.780
3300.00   92.320
2978.00   41.710
2938.00   41.330
2915.00   35.255
2518.00   96.450
2431.00   96.140
2361.00   91.640
2064.00   97.580
1716.00   43.175
1668.00   3.170
1664.00   36.630
1620.00   4.105
1449.00   17.580
1361.00   22.255
1358.00   15.440
1265.00   88.755
1220.00   12.665
1166.00   14.010
1088.00   88.035
1019.00   54.840
965.00   24.170
901.00   76.585
823.00   48.645
822.00   12.175

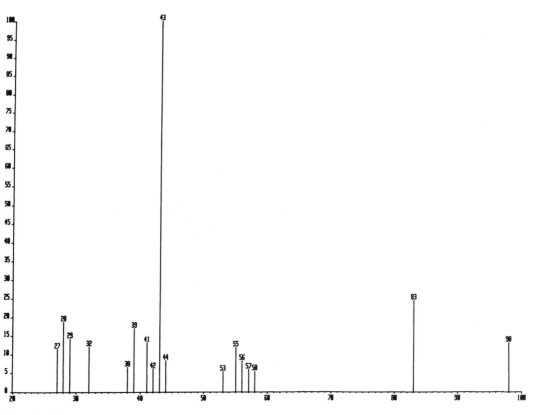

Sample 12.20

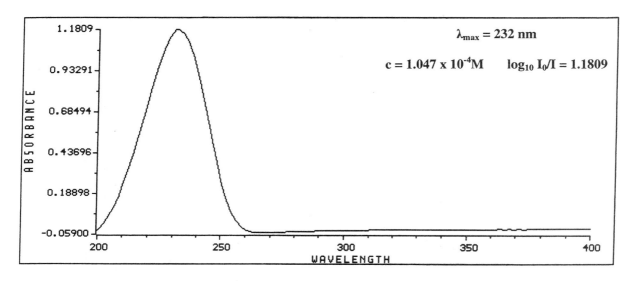

λ_{max} = 232 nm

c = 1.047 x 10^{-4}M   log$_{10}$ I$_0$/I = 1.1809

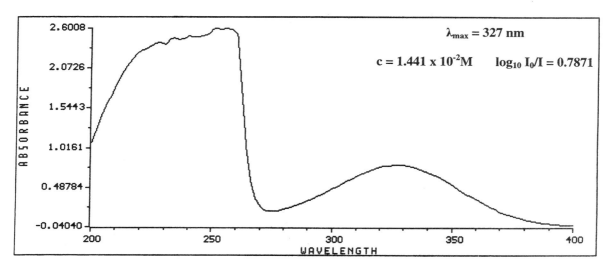

λ_{max} = 327 nm

c = 1.441 x 10^{-2}M   log$_{10}$ I$_0$/I = 0.7871

Sample 12.20 *Continued*

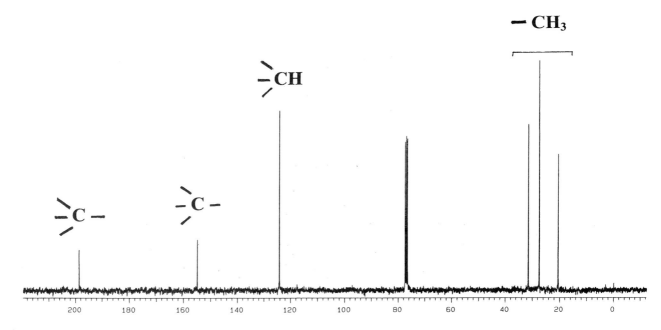

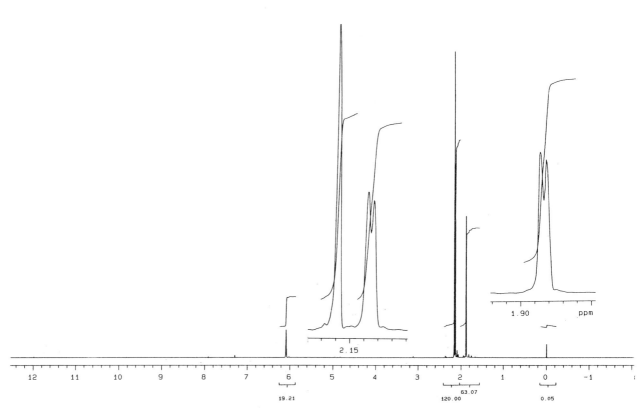

Sample 12.20 *Continued*

CHAPTER 13

Difficult Problems in Interpreting Spectra

The spectra in this exercise are more difficult to interpret than those in the previous chapter. This is not the result of supplying spectra of cyclic and rigid systems—you need to go more deeply into ^{1}H NMR spectroscopy to solve those. These spectra are more difficult to solve either because they are complex in the sense of having a number of functional groups, or because they come into the category which Professor Abraham once described as 'deceptively simple'. Three of the samples have NMR spectra which could hardly be simpler, but it does not make them easy to solve. You need to remain aware of the effects of symmetry in considering these spectra, but above all you need some imagination.

The spectra of 10 samples are provided, some easier than others. All can be solved from the spectra given—one or two of the spectra in fact are hardly necessary, which is exactly the situation you will find when working at the bench.

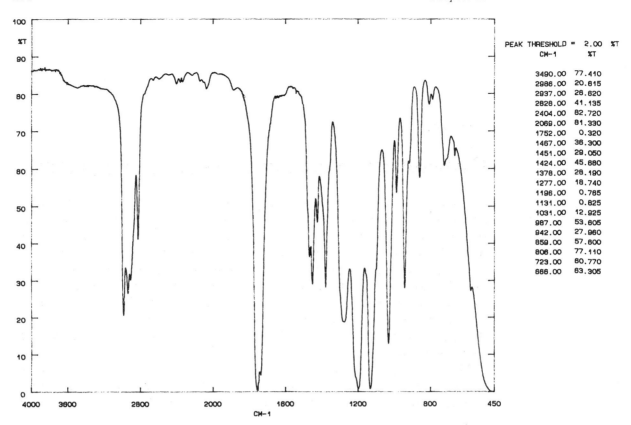

PEAK THRESHOLD = 2.00 %T

CM-1	%T
3490.00	77.410
2986.00	20.615
2937.00	28.620
2828.00	41.135
2404.00	82.720
2069.00	81.330
1752.00	0.320
1467.00	38.300
1451.00	29.050
1424.00	45.680
1378.00	28.190
1277.00	18.740
1196.00	0.785
1131.00	0.825
1031.00	12.925
987.00	53.605
942.00	27.960
859.00	57.600
806.00	77.110
723.00	60.770
666.00	63.305

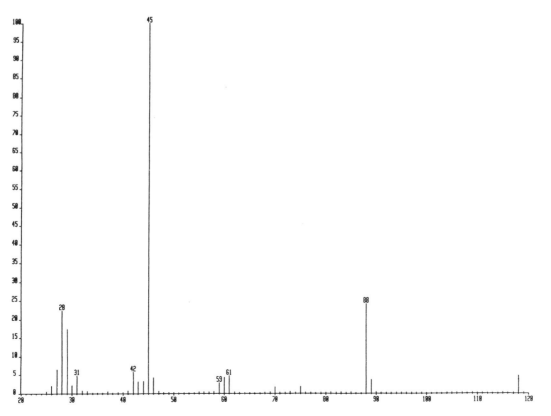

Sample 13.1

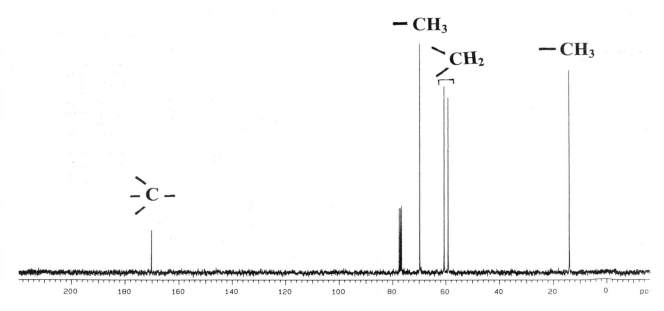

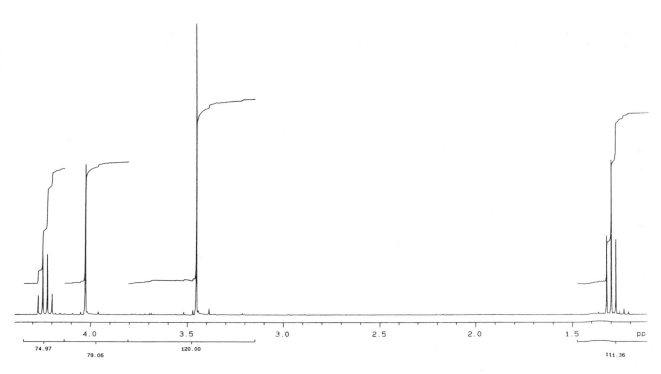

Sample 13.1 *Continued*

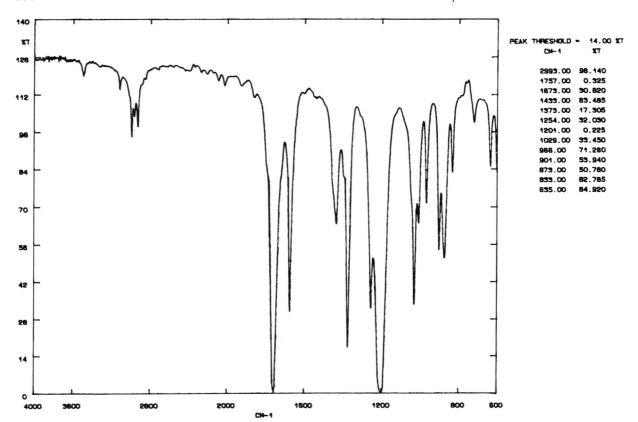

PEAK THRESHOLD =   14.00 %T
        CM-1   %T

     2993.00   96.140
     1757.00   0.325
     1873.00   30.820
     1433.00   83.485
     1373.00   17.305
     1254.00   32.030
     1201.00   0.225
     1029.00   33.450
      988.00   71.280
      901.00   53.940
      873.00   50.780
      833.00   82.785
      635.00   84.920

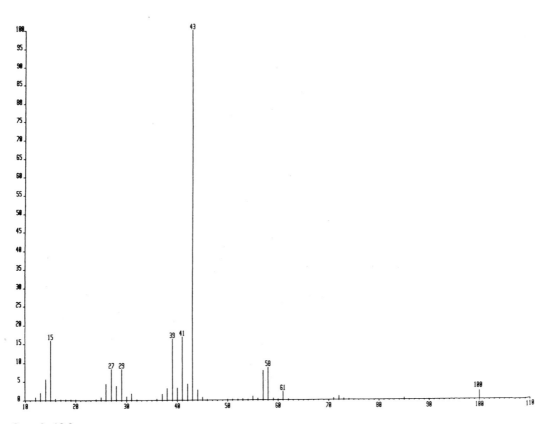

Sample 13.2

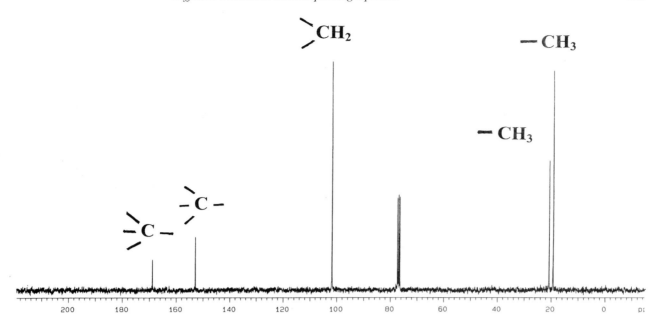

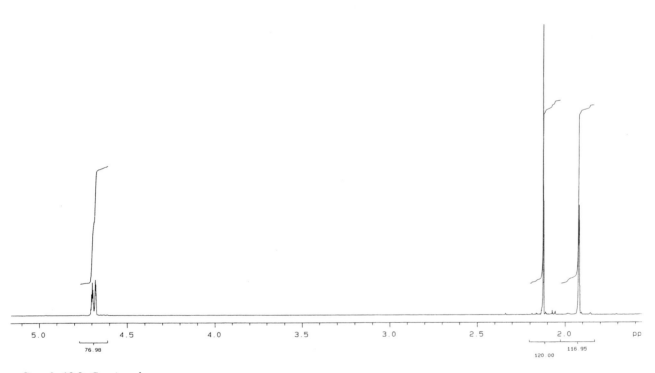

Sample 13.2 *Continued*

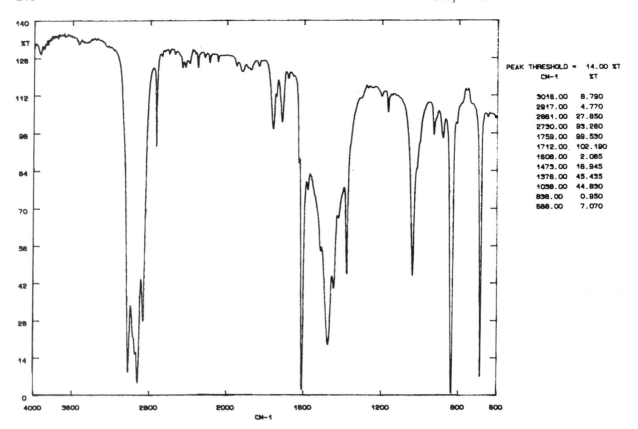

PEAK THRESHOLD = 14.00 %T

CM-1	%T
3016.00	8.790
2917.00	4.770
2861.00	27.850
2730.00	93.280
1759.00	99.530
1712.00	102.190
1608.00	2.085
1473.00	18.945
1378.00	45.435
1058.00	44.830
838.00	0.950
688.00	7.070

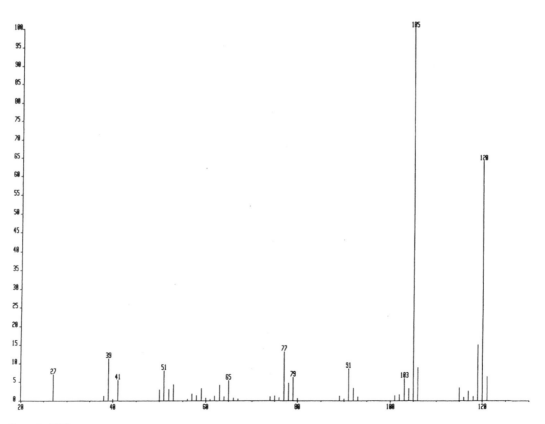

Sample 13.3

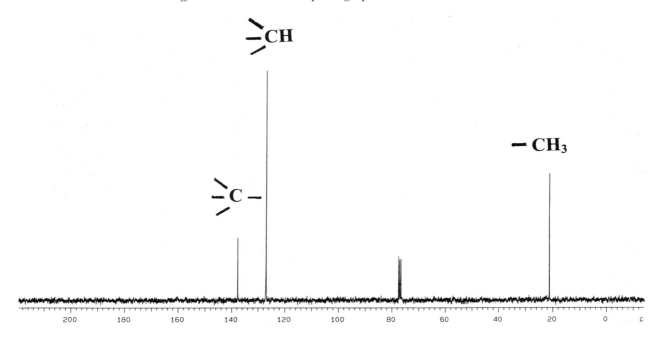

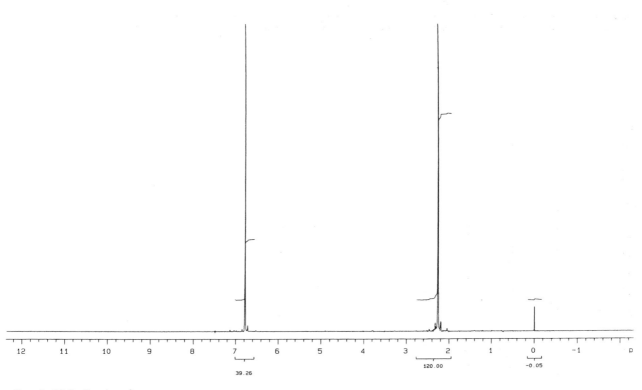

Sample 13.3 *Continued*

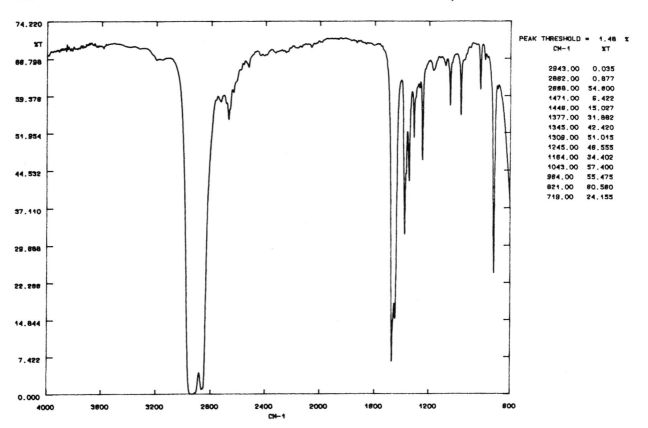

PEAK THRESHOLD = 1.48 %

CM-1	%T
2943.00	0.035
2862.00	0.877
2868.00	54.600
1471.00	6.422
1446.00	15.027
1377.00	31.882
1345.00	42.420
1309.00	51.015
1245.00	46.555
1164.00	34.402
1043.00	57.400
984.00	55.475
821.00	60.580
719.00	24.155

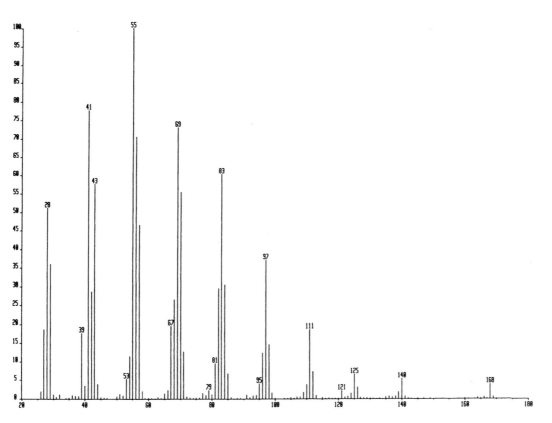

Sample 13.4

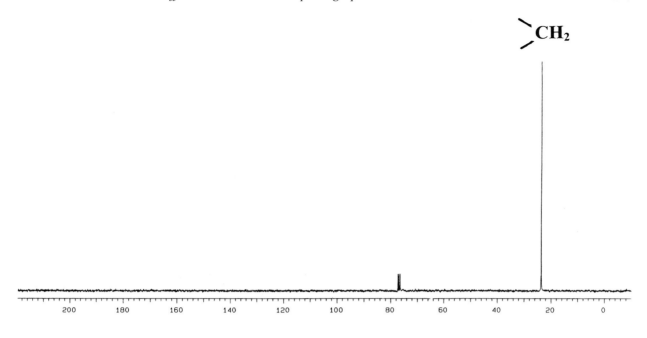

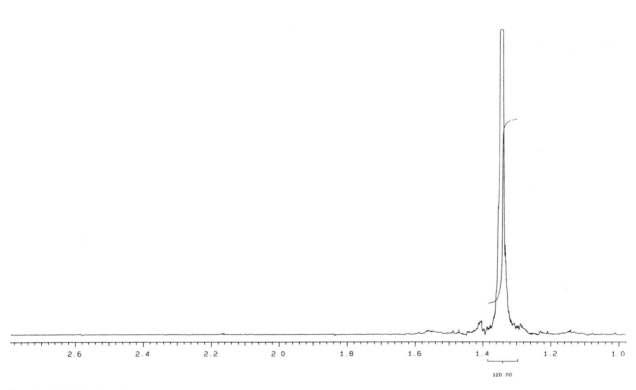

120.00

Sample 13.4 *Continued*

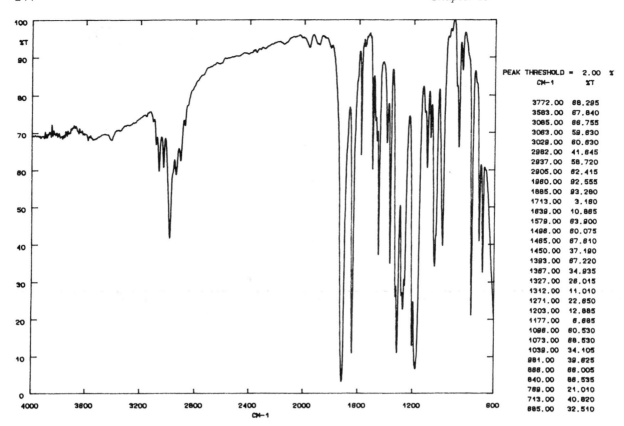

PEAK THRESHOLD = 2.00 %

CM-1	%T
3772.00	68.295
3583.00	87.840
3085.00	66.755
3063.00	59.630
3029.00	60.630
2982.00	41.645
2937.00	58.720
2905.00	62.415
1960.00	92.555
1885.00	93.280
1713.00	3.180
1639.00	10.865
1579.00	63.900
1496.00	60.075
1465.00	87.610
1450.00	37.190
1393.00	87.220
1367.00	34.935
1327.00	26.015
1312.00	11.010
1271.00	22.650
1203.00	12.885
1177.00	6.685
1096.00	60.530
1073.00	66.530
1039.00	34.105
981.00	39.625
866.00	66.005
840.00	66.535
769.00	21.010
713.00	40.820
665.00	32.510

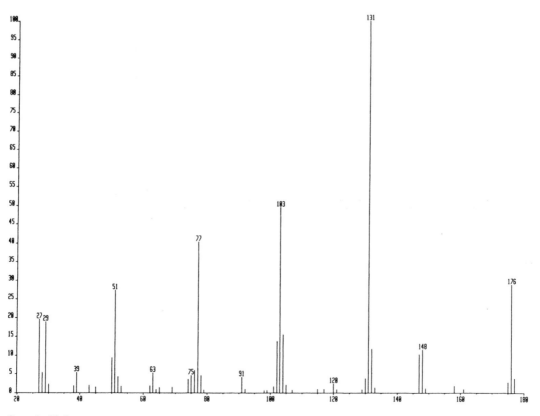

Sample 13.5

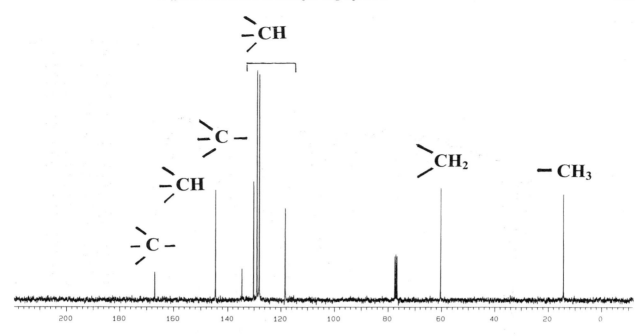

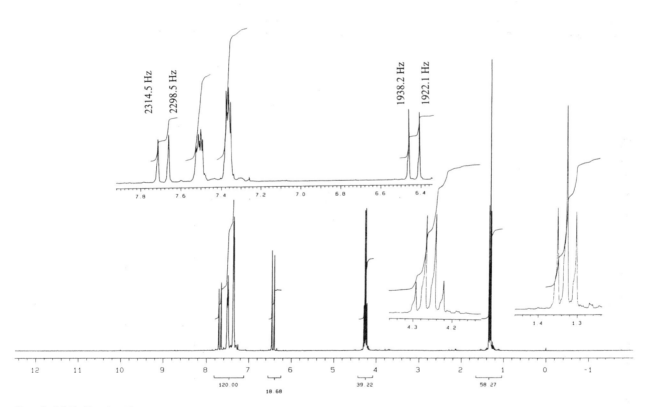

Sample 13.5 *Continued*

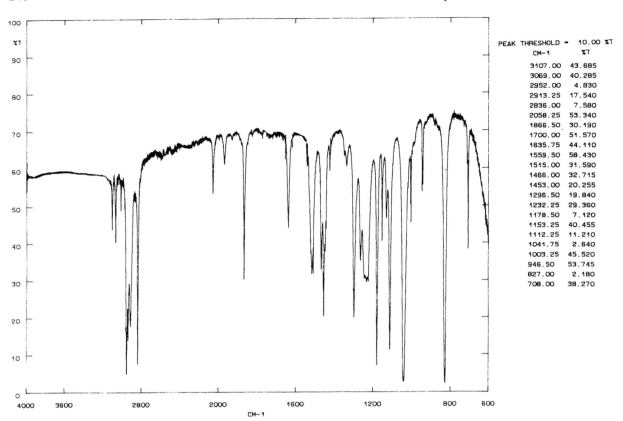

PEAK THRESHOLD = 10.00 %T

CM-1	%T
3107.00	43.685
3069.00	40.285
2952.00	4.830
2913.25	17.540
2836.00	7.580
2058.25	53.340
1866.50	30.190
1700.00	51.570
1635.75	44.110
1559.50	58.430
1515.00	31.590
1466.00	32.715
1453.00	20.255
1296.50	19.840
1232.25	29.360
1178.50	7.120
1153.25	40.455
1112.25	11.210
1041.75	2.640
1003.25	45.520
946.50	53.745
827.00	2.180
708.00	38.270

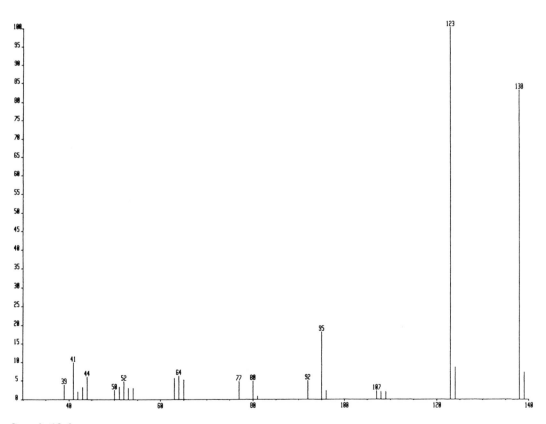

Sample 13.6

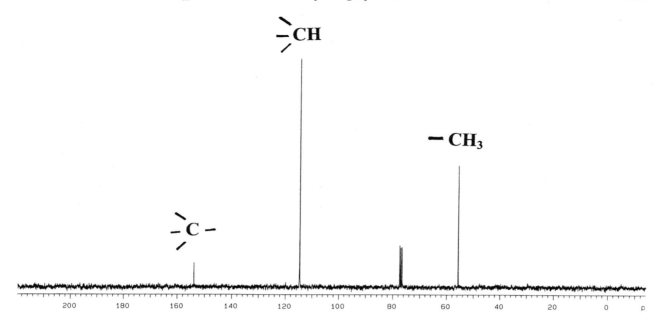

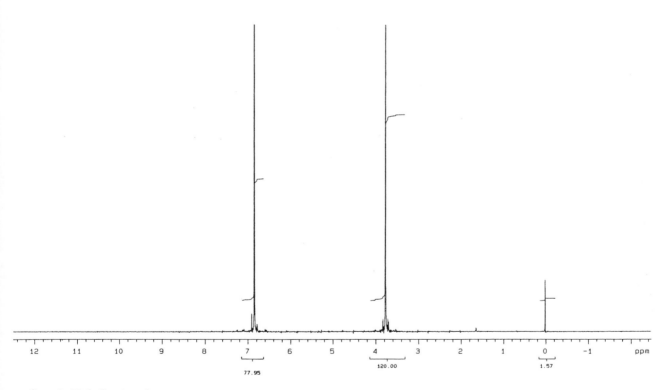

Sample 13.6 *Continued*

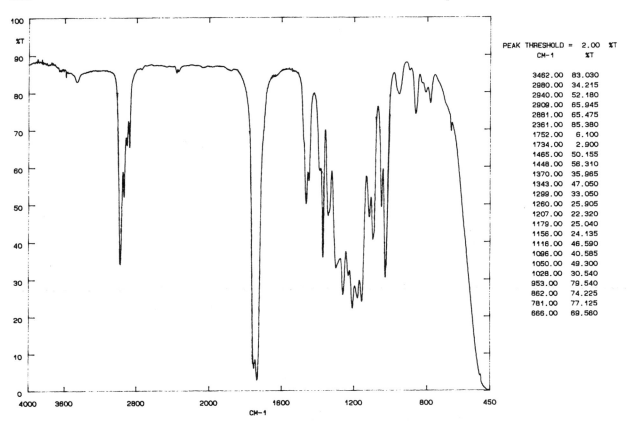

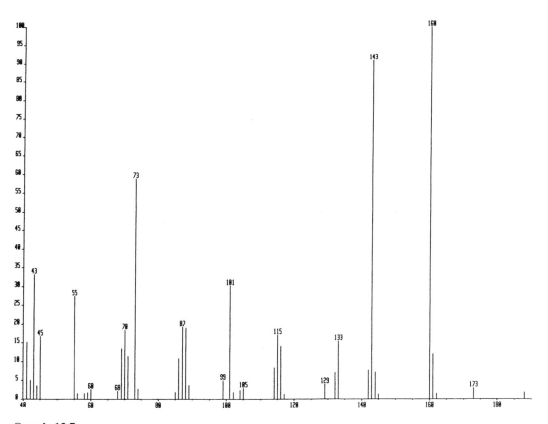

Sample 13.7

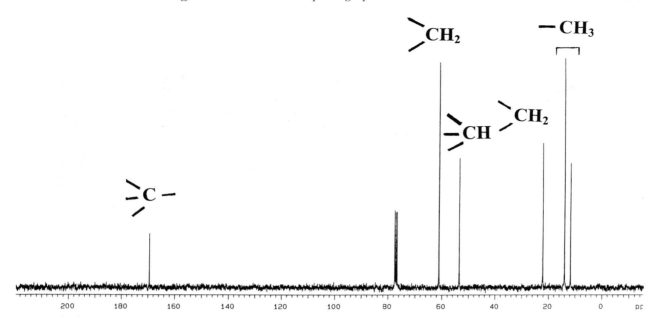

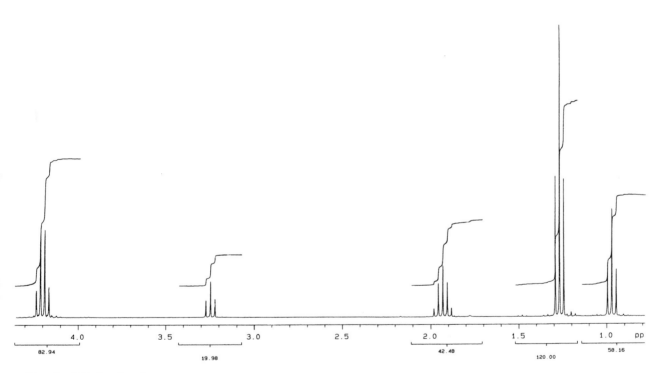

Sample 13.7 *Continued*

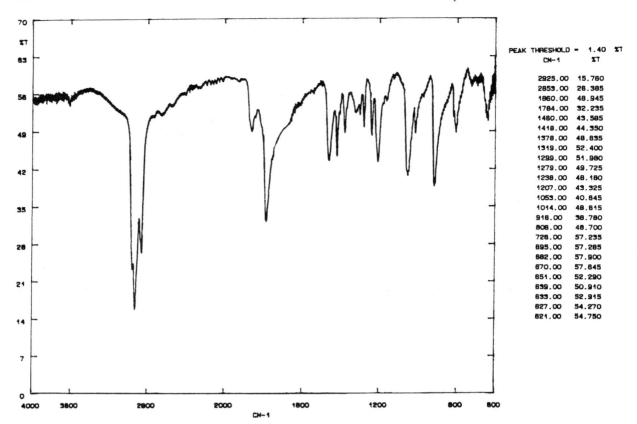

PEAK THRESHOLD =   1.40   %T

CM-1	%T
2925.00	15.760
2853.00	26.385
1860.00	48.945
1784.00	32.235
1480.00	43.585
1418.00	44.350
1378.00	48.635
1319.00	52.400
1299.00	51.980
1279.00	49.725
1238.00	48.180
1207.00	43.325
1053.00	40.845
1014.00	48.815
916.00	38.780
806.00	48.700
726.00	57.235
695.00	57.285
682.00	57.900
670.00	57.845
651.00	52.290
639.00	50.910
633.00	52.915
627.00	54.270
621.00	54.750

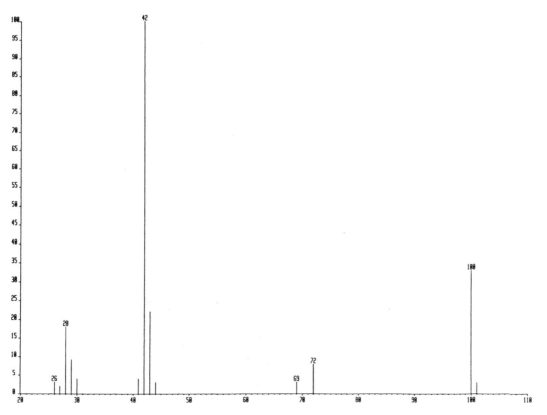

Sample 13.8

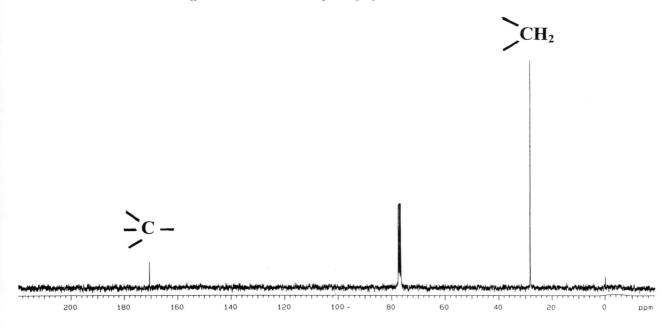

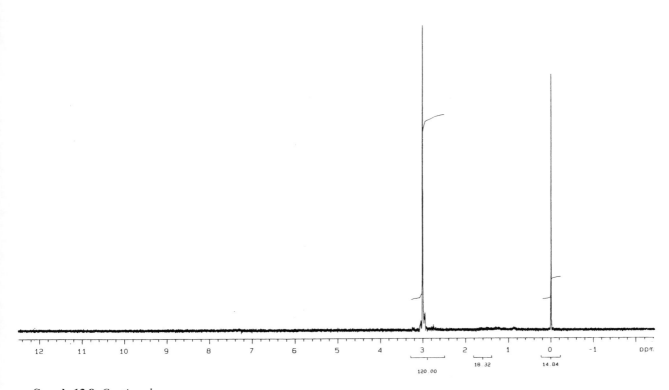

Sample 13.8 *Continued*

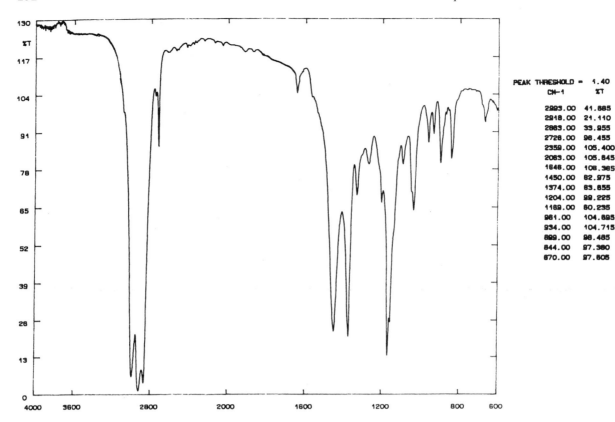

PEAK THRESHOLD =   1.40 %T
     CM-1   %T

     2993.00   41.885
     2918.00   21.110
     2863.00   33.955
     2726.00   98.455
     2359.00   105.400
     2083.00   105.845
     1846.00   108.385
     1450.00   82.975
     1374.00   83.855
     1204.00   99.225
     1189.00   80.235
      981.00   104.895
      934.00   104.715
      889.00   98.485
      844.00   97.380
      670.00   97.805

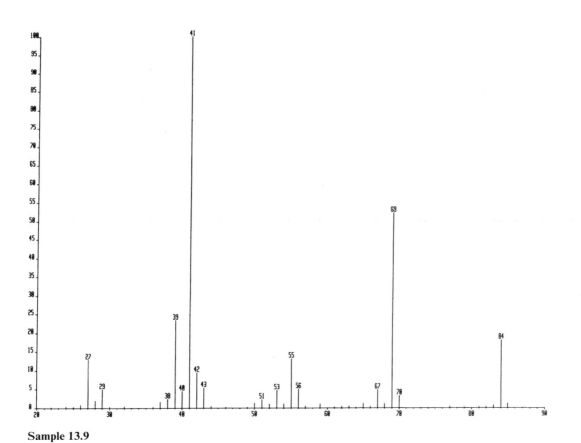

Sample 13.9

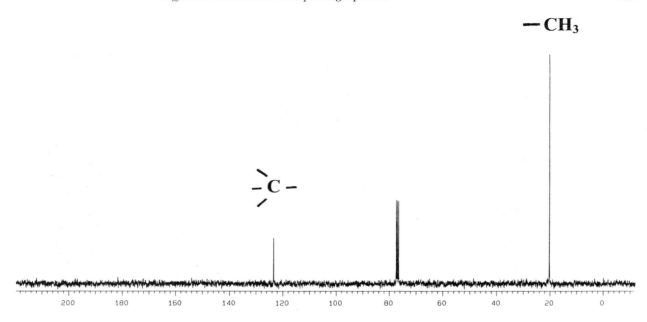

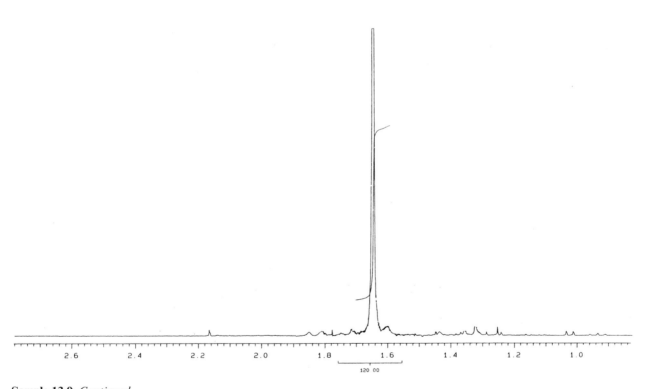

Sample 13.9 *Continued*

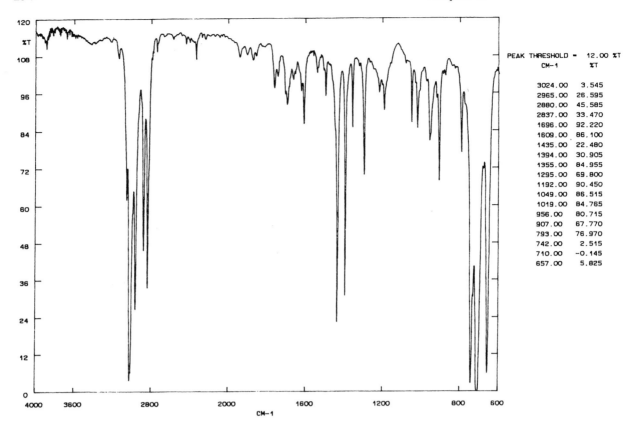

PEAK THRESHOLD = 12.00 %T

CM-1	%T
3024.00	3.545
2965.00	26.595
2880.00	45.585
2837.00	33.470
1696.00	92.220
1609.00	86.100
1435.00	22.480
1394.00	30.905
1355.00	84.955
1295.00	69.800
1192.00	90.450
1049.00	86.515
1019.00	84.765
956.00	80.715
907.00	67.770
793.00	76.970
742.00	2.515
710.00	-0.145
657.00	5.825

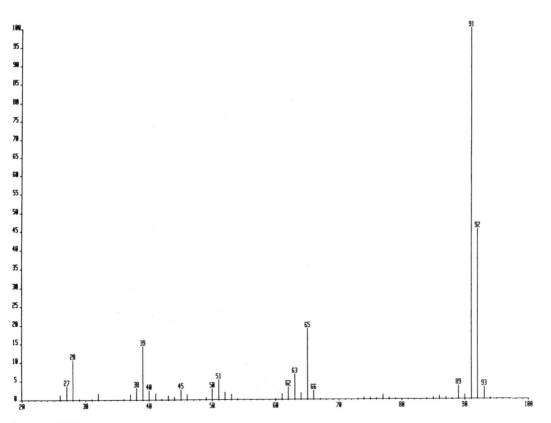

Sample 13.10

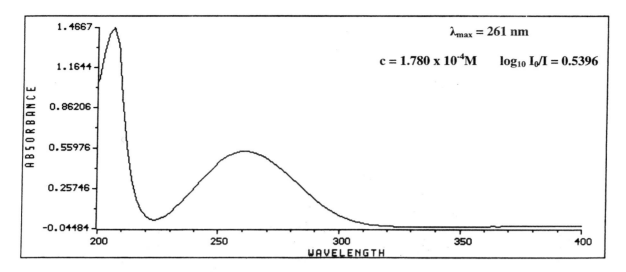

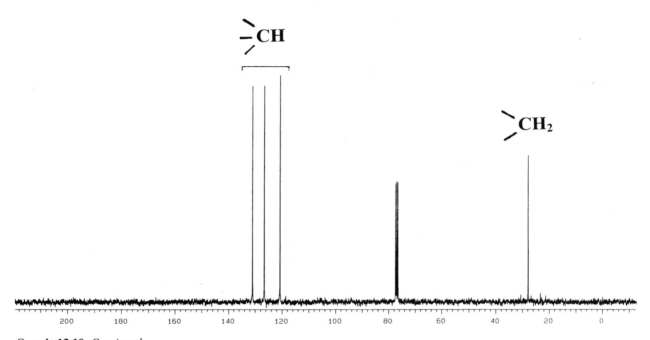

Sample 13.10 *Continued*

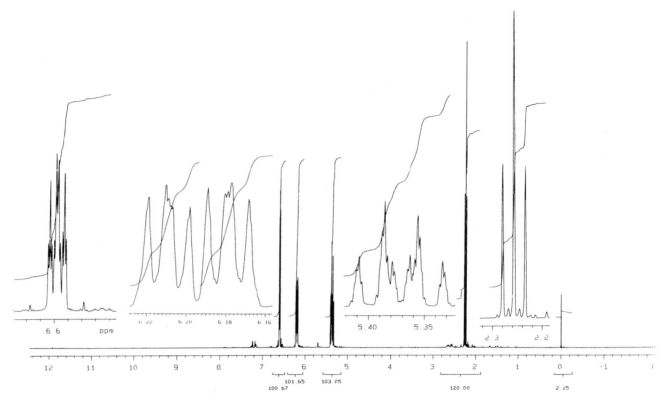

Sample 13.10 *Continued*

CHAPTER 14

Answers to Problems

Answers to Chapter 1 Problems

1. Alcohol
2. Ester of saturated acid
3. Primary amine
4. Saturated carboxylic acid
5. Saturated ketone
6. Alkene
7. Anhydride of saturated carboxylic acid
8. Alkynol
9. Ester of unsaturated acid
10. Aromatic carboxylic acid
11. Nitrile
12. Chloride of saturated acid
13. Secondary amide
14. Aromatic ketone
15. Saturated aldehyde
16. Anhydride of aromatic carboxylic acid
17. Chloride of aromatic acid
18. Aromatic aldehyde
19. Unsaturated ketone
20. Tertiary amide

Answers to Chapter 2 Problems

1. Phenylmethyl bromide (benzyl bromide)
2. Iodoethane
3. Hexane
4. Butanoic acid (butyric acid)
5. 2-Chloropropane
6. Bromoethane
7. Butylbenzene
8. Ethyl butanoate (ethyl butyrate)
9. 2-Aminopropane
10. Phenylmethyl chloride (benzyl chloride)
11. Methyl butanoate (methyl butyrate)
12. Butan-2-ol (*sec*-butyl alcohol)
13. 2-Chloroethylbenzene
14. Butanal (butyraldehyde)
15. Butan-2-amine (*sec*-butylamine)
16. 1,2-Dibromoethane
17. 1,1-Dibromoethane

18. Butanamide (butyramide)
19. (2-Aminoethyl)benzene
20. Hexan-2-one

Answers to Chapter 3 Problems

1. Propan-2-one (acetone)
2. Ethanol
3. 2-Bromoethanol
4. Acetonitrile
5. Benzaldehyde
6. Ethanoic acid anhydride (acetic anhydride)
7. Ethanoic acid (acetic acid)
8. Methyl acetate
9. Chloroacetonitrile
10. Pentan-3-one

Answers to Chapter 4 Problems

1. Saturated ketone
2. Unsaturated ketone
3. Unsaturated ketone
4. Unsaturated ketone
5. Triene

Answers to Chapter 5 Problems

1. Ethanol
2. Propanoic acid (propionic acid)
3. Propan-2-ol (isopropyl alcohol)
4. Methyl benzoate
5. 2-Methylbutane
6. Cyclohexene
7. 2-Phenylethanol
8. *N*,*N*-Diethylphenylamine (diethylaniline)
9. Di-*n*-butyl ether
10. Propylbenzene

Answers to Chapter 6 Problems

1. Propan-2-yl acetate (isopropyl acetate)
2. Butanenitrile (butyronitrile, $CH_3CH_2CH_2CN$)
3. Phenyl acetate
4. Butane-2,3-diol
5. Prop-2-enyl acetate (allyl acetate, $CH_2=CHCH_2OCOCH_3$)
6. Benzoyl cyanide (PhCOCN)
7. 3,3-Dimethylbutanal (3,3-dimethlybutyraldehyde)
8. Hexanedicarboxylic acid (adipic acid)
9. 2,2-Dimethylpropane-1,3-diol
10. 2-Ethylbutan-1-ol

Answers to Chapter 7 Problems

1. 1,2-Dimethoxyethane

2. 1-Bromobutane
3. 1,1-Dimethylethylbenzene (*tert*-butylbenzene)
4. Octanal
5. Phenylmethanol (benzyl alcohol)
6. Heptanoic acid
7. 1,1-Dimethylethyl iodide (*tert*-butyl iodide)
8. Bis(1-methylethyl) ether (diisopropyl ether)
9. Cyclohexanone
10. Oct-1-ene

Answers to Chapter 8 Problems

1. Pentan-2-ol
2. 1,6-Dibromohexane
3. 3-Bromopropanenitrile (3-bromopropionitrile, $BrCH_2CH_2CN$)
4. 1-Phenylpropan-1-one (propiophenone)
5. Butan-1-amine (*n*-butylamine)
6. Cycloheptanone
7. Octanoic acid
8. Methyl prop-2-eneoate (methyl acrylate, $CH_2{=}CHCOOCH_3$)
9. Methylenecyclohexane
10. Methyl hexanoate
11. 4-Methylpentan-1-ol
12. 1-Methylcyclohexanol
13. Hex-1-en-3-ol
14. 3-Chloropropan-1-ol
15. Heptan-2-one
16. 4-Methoxybenzaldehyde
17. 1-Methylcyclohexene
18. But-3-en-2-one (methyl vinyl ketone)
19. 2-Methylpentanoic acid
20. Cyclohex-2-en-1-one

Answers to Chapter 9 Problems

1. *N,N*-Diethylphenylamine (diethylaniline)
2. Propan-2-yl acetate (isopropyl acetate)
3. 2-Bromoethanol
4. 2-Phenylethanol
5. *cis*-3-Chloroprop-2-enoic acid (*cis*-chloroacrylic acid)
6. 4-Methoxybenzaldehyde
7. 3-Bromopropanenitrile (3-bromopropionitrile, $BrCH_2CH_2CN$)
8. *trans*-3-Chloroprop-2-enoic acid (*trans*-chloroacrylic acid*)*
9. 1-Phenylpropan-1-one (propiophenone)
10. 3-Nitrobenzoic acid

Answers to Chapter 10 Problems

1. 2-Methylpropanenitrile (isobutyronitrile)
2. 2-Methylpropanoic anhydride (isobutyric anhydride)
3. Diethylamine
4. Pentan-3-one
5. 3-Phenylpropanal (3-phenylacetaldehyde)
6. Ethyl phenylacetate

7. Propanoic acid (propionic acid)
8. Butanenitrile (butyronitrile)
9. Pentan-2-one
10. 1-Methylethyl benzoate (isopropyl benzoate)

Answers to Chapter 11 Problems

1. Diethyl ether
2. 1-Iodobutane
3. Ethylbenzene
4. 2-Chloropropane
5. 1,3-Dichloropropane
6. 1-Methoxy-4-methylbenzene (*p*-methylanisole)
7. Diethyl dipropanoate (diethyl malonate)
8. Di-*n*-propyl ether
9. 2-Methylbutane (isopentane)
10. 2,2-Dimethylpropane-1,3-diol

Answers to Chapter 12 Problems

1. Butan-2-one
2. 2-Methylpropan-1-yl acetate (isobutyl acetate)
3. Propanenitrile (propionitrile, CH_3CH_2CN)
4. 1-Phenylethanone (acetophenone)
5. Prop-2-yn-1-ol (propargyl alcohol)
6. 3-Methylbutan-2-one
7. Tripropylamine
8. Heptan-3-one
9. Butylbenzene
10. 4-Bromobenzaldehyde
11. Diethyl ethanedioate (diethyl oxalate)
12. 4-Methylpentan-2-one
13. *N,N*-Diethylformamide
14. Methyl 2,2-dimethylpropanoate
15. 1-(4-Bromophenyl)ethanone (4-bromoacetophenone)
16. 2-Methylpropylbenzene (isobutylbenzene)
17. Ethyl cyanoacetate
18. Diethyl butanedioate (diethyl succinate)
19. 4-Bromobutanenitrile (4-bromobutyronitrile)
20. 4-Methylpent-3-en-2-one (mesityl oxide)

Answers to Chapter 13 Problems

1. Ethyl methoxyacetate
2. Propen-2-yl acetate (isopropenyl acetate)
3. 1,3,5-Trimethylbenzene
4. Cyclododecane
5. Ethyl *trans*-3-phenylprop-2-enoate (*trans*-ethyl cinnamate)
6. 1,4-Dimethoxybenzene
7. Diethyl ethylpropanedioate (diethyl ethylmalonate)
8. Butanedioic anhydride (succinic anhydride)
9. 2,3-Dimethylbut-2-ene
10. Cycloheptatriene

Subject Index